Workshop technology

Part 2

# Workshop technology

## Part 2

SI UNITS

Dr W. A. J. CHAPMAN

MSc(Eng), FIMechE, Hon FIProdE

Routledge
Taylor & Francis Group

LONDON AND NEW YORK

First published 1946 by Edward Arnold (Publishers) Limited

*Second Edition 1954*
*Third Edition 1963*
*Fourth Edition 1972*

This edition published 2013 by Routledge
4 Park Square, Milton Park, Abingdon, Oxon OX14 4RN
605 Third Avenue, New York, NY 10158

*Routledge is an imprint of the Taylor & Francis Group, an informa business*

© W. A. J. Chapman 1972

ISBN 13: 978-0-415-50304-4 (pbk)

# Preface to first edition

The First Part of *Workshop Technology* has made so many friends that I have every confidence in offering this, its second volume. Here will be found all the aspects of machine-tool work that the reader should require for his basic knowledge of workshop engineering, together with chapters dealing with the elements of accuracy, measurement and gauging. The two parts together provide a complete introduction to the technique of the workshop itself and contain elements of all the knowledge a student should require whilst serving an apprenticeship, or learning period in the shops. Although primarily written in detail for the inhabitants of the workshop, the general terms of the text include nothing but important fundamentals in the knowledge of any mechanical engineer, so that its reading should be of help and interest to any student of engineering.

The text covers most of the work necessary for the City and Guilds Intermediate and Final Examinations in Machine Shop Engineering, and the Ordinary National Certificate in Workshop Technology. Students taking this subject in Higher Certificates should also find the books helpful with parts of their syllabus. A thorough insight into the elements of work-shop technology requires a knowledge of many related and contributory processes. In making these two volumes a fairly complete guide to the workshop itself it has been necessary to plan the additional text into a third part which I hope to complete in due course.

I have again received help of great value from various firms and institutions, and a name appended to a diagram reproduced often represents assistance much in excess of the mere permission to use the illustration. I should like to offer my sincere thanks for such co-operation, and assure those concerned that if the book fulfils the purpose for which it has been written, their trouble will have been worth while. Finally, I must again compliment the Publishers, and their draughtsmen, on the quality of their work in the face of such difficult circumstances.

W.A.J.C.

*Stafford* 1945

# Preface to fourth edition

This book with its companion, *Workshop Technology, Part 1,* continues to play its part in the education of workshop engineers. Since the last revision, seven years ago, considerable progress has been made towards the recognition of our need to establish craft, in the workshop, as something worthwhile, and as an asset to our national well-being. Two important advances have been the establishment of the Industrial Training Boards and the clarification of the functions of craftsmen and technicians. Having now set up in the ways and means for producing our greatly needed supply of what I still prefer to call 'workshop engineers', it is now possible to look to the future with a greater degree of confidence.

The purpose of this present revision is to convert the text to the SI metric system of units and conventions and to bring the book up to date in its ideas and illustrations. The two books are now so well known that it seems invidious to state that they will prepare for this or that course of study or examination. I feel it sufficient to say that they will be useful textbooks for any system of education where technicians and craftsmen are concerned. In the past they have been used as much by readers outside the workshop as by those on the shop floor. I hope that they might still perform this useful service.

Once again I should like to acknowledge my appreciation for the help I have received from numerous firms and individuals. This time, with the changeover to SI it has been necessary to worry these people rather more than usual. The Publishers, as usual, have extended their customary encouragement and enthusiasm for the revisions and deserve my grateful thanks.

W.A.J.C.

*Hatfield* 1972

# Contents

**Appendix**

# 1 Accuracy, inter-changeability, gauging

Whatever may be the reader's work, if it has any connection with engineering construction, he must on many occasions have need to apply the fundamental rules of accuracy. In the workshop these rules are an essential part of the technique, and when the problem of checking surfaces arises the workshop engineer needs all the ingenuity and skill his mind and hands can muster. In addition, the action of checking work, whether his own or that of someone else, places him in the position of a judge, and for this reason he must exercise extreme honesty of mind and purpose, a process which, if the work is his own, may sometimes cause misgivings! The use of measuring instruments involves a delicacy of manipulation and touch which can only be acquired from experience. The inspection of accurate work is a job which cannot be hurried, and to make haste slowly is a lesson which is not often learned until after many delusions and disappointments. Let the reader be not discouraged, however, but let him remember that the most experienced of us have passed through the experiences he may sometimes find to be a test for his patience.

## Accuracy

A satisfactory component must have its important surfaces true with respect to form, relationship and dimensions. Trueness to form means that cylinders will be round, parallel and straight; flats will be flat, etc. Trueness to relationship implies that surfaces which should be perpendicular will be within a close approach to 90°, and so on. The degree to which perfection in accuracy is attained and achieved will depend, of course, on several factors, the chief of which are as follows:

1. **The grade of the product.** We look for and expect adherence to perfection in a Rolls Royce car than in one made to a lower price under mass-produced conditions.
2. **The classification of the product.** Greater accuracy is necessary in a

high-efficiency aero engine than in one of a lower type operating under less arduous conditions of service. Slow-moving agricultural machinery need not be made to such fine tolerances as high-speed reduction gearing and so on.

**3. The function of the surfaces concerned.** The surfaces making up any component serve one or more of the following functions when the component is in use:

(a) Fit and work (move) against a corresponding surface on a mating component, often whilst carrying a load.

(b) Fit against a mating surface without any relative movement taking place.

(c) Serve as one of the surfaces necessary to make up the form of the component, and when the component is in use either clearing other surfaces or mating components, or being just an open, joining face of material.

In general, the degree of accuracy necessary is highest in (a) and lowest in (c), although, as we shall discuss presently, condition (b) may demand higher accuracy than (a) on certain components. The conditions under (c) include the various clearance and non-fitting surfaces found on every component. These may vary from such examples as (i) working clearance on a piston between the rings (a few hundredths), or (ii) a bolt hole (clearance from 0·20 mm upwards) to the rough unmachined surface of a casting which only 'fits' the atmosphere ('atmospheric fit').

**4. Conditions imposed by, and dependent on, operating factors.** For most components there are conditions which influence the accuracy required, or the possible departures from it. As an example we might consider the head tube of a bicycle frame—the tube through which the spindle tube of the front forks passes. The only machined faces here are the spherical seatings upon which fit the head ball races. The inside of the tube is clearance, and the outside is an atmospheric fit, a good finish being necessary, however, to provide an attractive enamelled surface. The conditions necessary for the spherical seatings at top and bottom may be summarised as follows:

(a) The spherical radius must be the same as that formed on the ball races or a snug fit will not be obtained, and movement may take place in service.

(b) Slight latitude is possible on the top diameter of the seating since the locknut will accommodate for slight variations in the up and down position of the ball races.

(c) The seating should be fairly well centralised in the rather thin tubing or weakness will be caused.

(d) The axis joining the centres of the spheres (top and bottom) of which the seatings form part, *must* be in line with the frame when viewed from behind. If this is not so the front and back wheels of the machine will not be in the same plane and the steering will be affected.

(e) The above axis when viewed from the side should be sloping as near as possible at the angle decided upon during the design of the machine. A slight deviation from the actual angle may be permissible without detrimental effects, but it must be remembered that a small variation over the length of the tube will have a relatively large effect on the position at which the wheel touches the ground.

**Synthesis of accuracy**

Before we pass on to the discussion of examples illustrating the points we have just set out it might be well to stress one important point. It is essential to start the work right, and as we proceed to observe the principles which will preserve accuracy. Let us consider one or two simple cases: The only way to achieve a perfect fit between a cylinder and a cylindrical hole is to have them both perfectly round, parallel and straight. They must, of course, have the necessary size relationships to give the fit we require, but we will discuss that later. If we started with a hole not round, or parallel, or with its sides not straight and tried to accommodate a shaft to it, we should never achieve our object. We might delude ourselves into thinking we had done so, but such a delusion would either be the result of our inexperience, or, if we knew better, would be equivalent to acting an untruth. If a true hole and shaft are obtained at the commencement nothing seems easier than making the necessary fit. A similar set of principles operates in the case of machining plane work such as the shaping up of a cube. If we commence by making one face truly flat, and proceed by making a second face truly square with the first we are able to build up the accuracies we are seeking in a systematic manner.

The contribution of a series of accurate intermediate steps in building up to some final desired result is illustrated in Fig. 1, where it is important that the face of the disc attached to the end of the shaft should be perpendicular to the base of the bracket and parallel with the tenon slot (A) machined in it. This result may be achieved automatically as follows, and no other method would be acceptable by any experienced workshop engineer.

1. Assuming the base to have been machined flat, and the tenon slot straight and parallel, the bracket would be set up so that when the boss is bored the axis of the hole will be perfectly at right angles to the tenon slot sides and parallel with the base of the bracket.

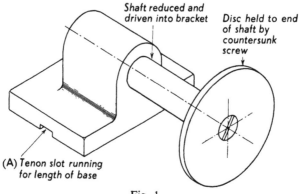

Fig. 1.

2. The bored hole must be perfectly round, parallel and straight.

3. The portion of the shaft which fits in the hole must be round, parallel, straight and a good fit in the hole.

4. The end of the shaft against which the disc fits must be flat (or slightly concave) and true with the fitting portion of the shaft. This can be achieved by facing it on the same *true* lathe centres, using the centre holes upon which the end was turned.

5. The faces of the disc must be parallel.

If these simple elements of accuracy are independently observed the final result desired will be achieved when the parts are assembled. The reader should notice that the diameter of shaft between the bracket and the disc plays no part in the desired alignment but merely serves as a carrier.

As a final example we might refer the reader back to a consideration of the suspension of the front wheel of a bicycle, and its alignment when viewed from front or rear. Assuming that the frame is not twisted and the back wheel to be truly centralised, the front wheel will be in alignment if the following alignments are correct:

1. Ball race seatings in the centre of the frame, and on an axis parallel to the frame, as previously discussed.

2. Ball races true, i.e. ball track concentric and parallel with spherical outside form.

3. Tubular spindle shank of forks straight, and with its axis in line with the centre line of the forks.

4. Axis of wheel spindle holes, or seatings, perpendicular to the common axis of the forks and shank.

5. Wheel on true spindle and bearings, wheel itself true and centralised between forks.

We hope, by these simple examples, that we have been able to convince the reader that the achievement of some desired alignment must be through the strict observance of simple fundamental rules. There is no other way, and any attempt to sidetrack the issue by unorthodox methods will surely lead to disillusionment and disaster. The principles are of such importance that we cannot stress them too strongly, and if the reader is to rise above the common level to take his place at the top of his profession he must keep them always before him. We hope, when dealing with machining methods, to offer suggestions for ensuring correctness in the most important fundamentals of accuracy.

**Conditions of accuracy**

It is the practice of many drawing offices when putting out drawings into the shops to ensure that they convey to the workshop engineer a complete and infallible set of working instructions. There are, however, probably as many if not more establishments in which the drawing when it reaches the shops, is incomplete, and leaves certain points in its interpretation to the knowledge and experience of the machinist or fitter on the job. Even when the drawing is supposed to be infallible, controversial issues often arise which depend for their solution on the intelligent application of knowledge and experience. There is no doubt that much unnecessary wastage is still caused by work being rejected due to conflicts of opinion as between the Inspection and the other departments responsible for the quality of a production. As a nation, our welfare and standard of living are so vitally concerned with our ability to produce more, and to avoid waste, that we should use every effort to ensure that whilst preserving the fundamental requirements of a design, we adopt a realistic outlook on small and unimportant details.

The knowledge and instincts necessary to be able to know which surfaces on a component must be correct, which may suffer from a certain amount of latitude, where a good finish is important, when an excess or shortage of metal can be tolerated and so on, will be mainly acquired by study, observation and experience. The sooner the reader can equip himself with such knowledge the better, because it will give him confidence from which will develop the initiative necessary to enable him to advance. To provide him with a starting point we will discuss briefly one or two simple examples.

Most of the components we make play a part in some kind of assembly and fit together with other components. When the whole thing is completed it must fulfil its function efficiently whether it be a mincing machine or an aircraft engine. If material and workmanship have been expended in excess of efficient requirements, time and money have been wasted. If it is inefficient because of the lack of care in manufacture, time and money have again been wasted. The design may, of course, be at fault, but we, as workshop engineers, are concerned with making the best of whatever design is supplied to us.

The first thing we should know is the purpose and use of the complete unit of which we may be making a part; whether it is an assembly jig, or a carburettor, or a gearbox and so on. Next comes a knowledge of the parts which go to make up the assembly, how they fit together and what duty each one plays in the working of the complete thing. In addition, we should know:

(a) What forces (if any) must be sustained and in which directions they act. From such knowledge we are able to judge the directions in which distortion and wear are likely to take place and to adjust the finish and fit accordingly.

(b) Which elements of the assembly are most vital to its efficient operation. In a plunger pump, for example, the fit of the plunger in the barrel is of supreme importance.

(c) Any special requirements which must be fulfilled by one or more of the components. For example, in making four feet to support an article it is more important that they should be uniform in height than that they should be particular to any dimension. The trueness of a mandril on its centres is of more importance than the accuracy of its diameters and so on.

We might, with advantage, complete this discussion by considering one or two actual examples.

Fig. 2 shows a sketch of a general purpose facing bar and cutter for use on a drilling machine. The bar would be made of a tough, medium carbon steel unhardened, the cutter from high-speed steel hardened, and the set pin could be a stock article, preferably with its head and point potash hardened. The taper, which fits the taper hole in the spindle of the machine, must be finish-turned or ground to a Morse taper gauge, leaving it projecting 6 mm from the end of the gauge. It is important that the finished taper fits the gauge for its whole length, otherwise the bar will not run true in the machine, and the loss of driving friction through the badly fitting taper might result in the tang being twisted off. The tang thickness should be central with the bar axis, and it should be kept well up to 16 mm. (There is 0·25 mm clearance for it in its mating slot.) The tang length, joining radius and 36 mm diameter relief are relatively unimportant, and with the length of the taper may vary within ±0·25 mm. The parallel portion of the bar should be true with the taper or the cutter, which is located from this portion, will not run truly. Its length and diameter are unimportant except in so far as the diameter forms a location on to which the cutter registers with the slot cut in its back edge. When the slot is made, its width should be

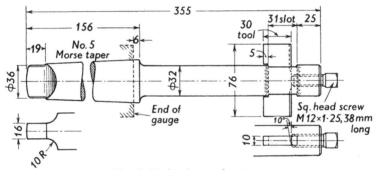

Fig. 2 Facing bar and cutter

parallel, a nice sliding fit for a 10 mm plug or slip gauge, and it should be central and parallel with the bar axis to within 0·10 mm. The end of the slot against which the cutter is pressed by the screw is important and should be flat, square with the slot sides and square with the bar axis, upon this last accuracy depending the truth of the face made by the cutter. The length of the slot and its distance from the bar end may be to within ±0·25 mm. Although the drawing shows the slot parallel with the taper tang their relative positions are immaterial and are shown this way merely for convenience of drawing. For the tapped hole and screw a good fit is necessary, and the end of the screw should be relieved as shown, to avoid the end thread being burred over, rendering the screw difficult to extract. The chief requirements for the cutter are that its two cutting edges should be straight, and both on the same straight line. This line should be parallel with the bottom of the slot with which the cutter presses against the end of the slot in the bar, where good contact should take place by accuracy in flatness and squareness at the bottom of the slot. The length and overall width of the cutter are unimportant (say, $\pm 0.25$ mm) and the cutting clearance should be within $\pm \frac{1}{4}°$.

Fig. 3 shows the end shield and bearing housing of a small electric motor. One of these is bolted to each end of the frame and they serve to carry the bearings for supporting the armature shaft in addition to carrying the brush gear, terminals, etc. The body and frame of the motor will be machined with an end face and male spigot for accommodating the outer face and spigot A of the end. Inside the body of the motor will be field coils on pole pieces arranged so that their ends will form a central bore slightly larger than the armature to be carried by these ends. It will be arranged that the central bore of the pole pieces is co-axial with the male spigots and square with the end faces of the frame, so that from such a starting point we have a clear conception of the requirements for the ends. The diameter of the spigot A is important as it fits a corresponding diameter on the frame and locates the end. The bore B is also important as regards size, because it

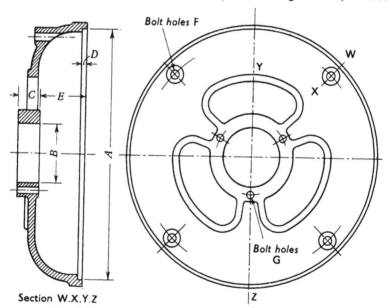

Fig. 3 End shield and bearing housing for electric motor

must provide a good fit for the outer race of a ball-bearing. It is very important also that this bore is concentric with the spigot A, and square with the end face of the shield where it bolts against the frame. An error in concentricity will cause the armature to be out of centre with the pole faces and an error in squareness will strain the shaft and bearings, leading to trouble and failure. The distances $C$ and $E$ are fairly important as they control the longitudinal position of the armature shaft and $E$ may control the position of the brush gear. Depth $D$ is not very important as some clearance is provided between the end of the spigot and this face. The bolt holes $F$ must be large enough to admit the bolts and must register with corresponding holes in the frame. Their centres, spacing and angular location, therefore, are important, and their diameter must provide clearance for the bolts. Holes $G$ in the bearing bosses are for the bolts which hold the bearing retainer caps. These have the same importance as the other bolt holes and must also have correct angular location with those.

### Interdependence of accuracy of size and form

As a third example we will consider the fitting lug and jaws shown at Fig. 4(a), which will provide us with an interesting case where one accuracy is related to another. For the outer faces of the lug to fit between the jaws of

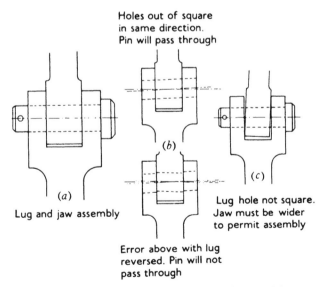

Fig. 4 Showing conditions of assembly of lug and jaw

the fork all the faces must be flat and parallel with the jaws one or two hundredths wider than the lug. If the pin is to pass through and fit the two holes in the jaws they must be round, parallel, co-axial and a push fit for the pin. For the pin to fit the hole in the lug the hole must be round, parallel and a push fit for the pin. The pin itself must be round, parallel and straight. Now let us consider the assembly of the three components. If all the above conditions are satisfied and the holes in the jaws and lug are *square* with the faces the whole will assemble and the pin will enter correctly. If by some chance the holes are not square and the error is the same in lug and jaws the pin will go through only when the other two elements are assembled in the direction which throws the errors in the same direction; reverse the lug and the pin will not pass through (Fig. 4(b)).

Now let us imagine that some deviation from squareness of the hole is to be tolerated and that it occurs on the lug. If the holes in the jaws are square and co-axial, the pin will not pass through the assembly now whichever way the parts are put together *unless* sufficient clearance is allowed either between the jaws, or in the fit of the pin, to permit the lug to tilt over sufficiently for the hole to line itself up, or for the pin to enter under the conditions of misalignment (Fig. 4(c)). Thus we see how the adherence to one set of conditions depends on another. If some deviation from hole squareness is allowed then a relative degree of slackness in the fit at the jaws is necessary to ensure that the components will assemble. This simple case is representative of many examples in practice where close and dependent relationships operate between the sizes and alignments of mating parts.

**Quantity production**

Our discussion so far has been limited to the general principles of accuracy which apply in equal measure to the production of one, or of a large number of components. There are, however, additional factors to be considered when production is on a large scale.

When only one or a small number of similar assemblies are being made, it is possible for the making up of the components to be effected more or less on an individual basis with particular points being attended to during machining and the fitter putting the finishing touches during assembly. For example, if a single assembly of Fig. 1 were required the bracket casting would first be shaped up on its base and would then pass on to have its boss bored and faced, either on a lathe or a boring machine. It would then be available for the turner who made the shaft to try in the shaft and assemble the same when the desired fit had been obtained. The same turner might have bored the bracket and would probably make the disc, put the screwed hole in the end of the shaft and finish the job. If the order were for half a dozen the procedure would be on similar lines, and, although production might not be quite such a one-man affair, it would still be possible for each shaft to be fitted to its own bracket.

When large numbers have to be made, however, a great change comes over the entire scheme of working. If the assembly we are considering were subject to such a demand that its production ran into hundreds, or thousands, it would become an article to be sold in a competitive market and its manufacture to a price would become a factor of much greater importance than when one, or only a few, were being made to special order. In addition to the question of economy in cost is that of producing this order along with others to a given time programme in one or more machine and assembly shops. The reader will readily appreciate that for an order of hundreds, it would not be expedient to allow one (or even several) turners to bore the brackets, then make the shafts to fit and then turn the discs, and so on. So that the brackets shall not wait whilst the shafts and disc are being made it must be possible, whilst the brackets are being processed on one machine, for the shafts and discs to be undergoing their machining operations on others. Such a method of working introduces the principle of the division of labour, a principle which has made possible great economies and high efficiency in quantity production. Under such conditions, the production of our assemblies would be carried out in one or more batches depending on the total quantity to be produced and on such factors as the periodic requirements of the market, etc. The bracket castings, the bar material for the shafts and the blanks for the discs, in sufficient quantity to produce a batch (say, 100), would be put out into the shop and the operations might take place as follows:

(*a*) Bracket.
> 1st Operation. Machine base and tenon slot.
> 2nd Operation. Machine faces of bosses (probably on a different machine from the 1st operation).
> 3rd Operation. Bore hole (on another machine).

(*b*) Shaft.
> 1st Operation. Rough turn.
> 2nd Operation. Finish turn (probably on another machine).
> 3rd Operation. Drill and tap hole in end.

(c)  Disc.

    1st  Operation. Face one side and bore hole.

    2nd Operation. Face other side and turn top (probably on another
        machine).

(d) Bracket, shaft and disc. Assemble.


Machining operations on all three components would probably be taking place at the same time and, from the analysis we have assumed, eight different machinists might be concerned with the machining. It is possible, furthermore, that the different components must be machined in different shops and even in different localities, so that the shafts will not see the brackets until they come together for assembly. When assembly does take place, any one of the hundred shafts must fit any one of the hundred brackets, and the same for the discs. We see, therefore, that quantity production introduces problems in connection with the mating together of the components of an assembly.


## Interchangeability

It will probably occur to the reader that uniformity is an essential condition of production of this type, and we must impress upon him that, in addition to the principles we have already discussed, uniformity is important where quantity production is concerned; in fact, it is the essence of cheap mass production. To what degree, however, can human skill, working to a set competitive time, maintain uniformity? We will assume that the shaft, where it fits the bracket, is 25 mm diameter, and that a driving fit is necessary. This means that the shaft must be about 0·02 mm greater than the hole, so that if all the holes are made 25·00 mm diameter, all the shafts should be 25·02 mm, and then any shaft will assemble correctly in any bracket. Given sufficient time, any turner could work to and maintain these sizes to within a close degree of accuracy, but there would still be small variations. When the work has to be done in a set competitive time it is only to be expected that the variations will be greater. This is recognised and a compromise effected in which certain variations are allowed in the sizes of the hole and the shaft.

Reverting to our example and assuming that a running fit is required it might be found that, provided the hole diameter exceeded the shaft by any amount from 0·02 mm to 0·06 mm, the fit would be satisfactory. (Naturally, the smaller clearance would give a closer fit than the larger, but either might meet the case.) In such a case, if the hole were made 25·00 mm, any shaft between 24·94 mm and 24·98 mm would be satisfactory for the fit required, but to allow a margin for variations in the hole as well we must reduce the margin on the shaft and give some to the hole. This could be effected by allowing the hole to vary between 25·00 mm and 25·02 mm, and the shaft between 24·96 mm and 24·98 mm. Then the combination of the largest shaft with the smallest hole would give a shaft 0·02 mm smaller, and the smallest shaft with the largest hole would give a shaft 0·06 mm smaller, which is what we require. When a system of this kind has been worked out, so that any one component will assemble correctly with any mating component, both being chosen at random, the method is called an *interchangeable system*.

### Elements of interchangeable systems

The hole and shaft we have just discussed are shown diagrammatically at Fig. 5, which also shows alternative methods of dimensioning, commonly used by drawing offices. The larger and smaller dimensions are called the *limits,* there being a *high limit* and a *low limit*. The difference between the high and low limits, which is the margin allowed for variations in·workmanship, is called the *tolerance*. If all the tolerance is allowed on one side

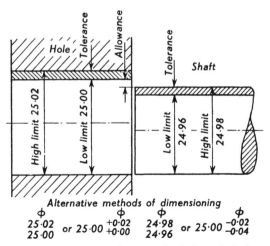

Fig. 5  Limits and tolerances for hole and shaft

of the nominal diameter $\left( \text{e.g. } 25 \cdot 00 \, {}^{+0 \cdot 02}_{+0 \cdot 00} \right)$ the system is said to be *unilateral* (uni = one), whilst if it is divided, some being allowed on either side of the nominal $\left( \text{e.g. } 25 \cdot 00 \, {}^{+0 \cdot 01}_{-0 \cdot 01} \right)$ the system is *bilateral* (bi = two). The bilateral system was allowed for in former limit systems, but the modern ISO system, described below, has abandoned it in favour of unilateral working.

An interchangeable system is generally called a *limit system* or a system of *limits and fits*.

For a system of limits to be really useful it must provide information from which the most usual kinds of engagement between two mating parts may be obtained. The four common classifications of fit are as follows:

(*a*) *Running fit*   A smooth, easy (but not loose) fit for the purpose of a moving bearing pair.

(*b*) *Push fit*   Can be assembled with light hand-pressure (locating plugs, dowels, etc.).

(*c*) *Driving or press fit*   Can be assembled with a hammer or by medium pressure. Gives a semi-permanent fit such as is necessary for a keyed pulley on a shaft.

(*d*) *Force fit*   Requires great pressure to assemble and gives a permanent fit. Used for wheels and hubs on shafts from which they are never likely to be removed.

These classifications may be further subdivided by making such fits as 'slack run', 'close run', 'light drive', 'heavy drive', etc.

To provide for various kinds of engagement between the shaft and hole there must be variations in the difference between their average sizes. For example, in a running fit the shaft is a small amount *less* in diameter than the hole (*clearance fit*), whilst to obtain a force fit the shaft must be a little *larger* than the hole (*interference fit*). Between the zones of interference and clearance there exists a range of fits which are neither one nor the other (e.g. a 'push' fit). These are called *transition fits*. The variation for the purpose of providing different classes of fit, is called the *allowance*. The various allowances for different fits may be obtained either (*a*) by keeping the hole constant and varying the shaft diameter to give the fit required, or (*b*) keeping the shafts constant and varying the hole. Keeping the hole constant is called the *hole basis*, whilst the reverse is the *shaft basis* method. All modern limit systems favour the hole basis, the chief reason being that most holes produced under quantity production conditions are machined with some form of reamer or fixed diameter cutter, whilst shafts are turned using tool boxes with easily adjusted tools. If the shaft basis system were employed very little benefit would be given to the production of shafts, but hole production would be much more complicated because for every type of fit, in each size, a different reamer would be necessary. By employing the hole basis one size reamer suffices for all the holes to any particular diameter. In a few circumstances it may be desirable to employ a shaft basis. A case in point is where a single shaft may have to accommodate a variety of accessories such as bearings, couplings, collars, etc. and it is desirable to maintain a constant diameter for the permanent shaft, and vary the fittings to suit. Some provision is made for shaft basis in the ISO limit system below.

**Limit systems**

The design of a limit system is not as simple as might be supposed. Consideration has to be given to the allowances suitable for the various classes of fit included, and to ensure that the desired fit will still be obtained under all possible variations of hole and shaft within the limits set out. The allowance for any fit will vary according to the diameter, and conditions for medium or low-class work will not be suitable for high-class products. If the limits are too close (small tolerance) a better control of the fit is possible, but production will be expensive. Wide limits (large tolerance) will cheapen production, at the same time introducing the risk that mating components which may happen to be on opposite extremes of size will not provide a satisfactory fit. To summarise, the limits and allowance will depend on the following:

(a) Nominal size (e.g. whether 25 mm, 100 mm, etc.).
(b) Class of fit required.
(c) Quality of product.

The limit system generally employed in this country is the British Standard System.

**The ISO system.** This system, set out in BS 4500: 1969, allows for 27 types of fit and 18 grades of tolerance, covering a size range of zero to 3150 mm. At first sight the provision of 27 fits with 18 grades of accuracy seems to offer a much wider selection than should ever be required but the provisions allow for everything from fine gauge and instrument work to the roughest form of production. Average workshop requirements may be met from a limited range of the holes and shafts and recommendations are made with which we shall deal below. In the system the 27 possible holes are designated by capital letters ABCDE . . . etc., and the shafts by small letters covering the same range. The 18 accuracy grades are covered by the numerals 01, 0, 1, 2, 3, . . . 16. For specifying any particular hole or shaft the rule is to write the letter followed by the numeral denoting the tolerance grade, e.g. H7 for a hole or f7 for a shaft. A fit involving these two elements is written H7—f7 or H7/f7.

For ordinary engineering practice the H holes and particularly H7, H8, H9 and H11 are recommended as being satisfactory for most purposes. Details of the limits for these holes over a diameter range of 6 mm to 250 mm are shown in Appendix 3.

The general trend for the shafts is that the range a to g have both limits less than the nominal size and tend to give clearance fits. The h shafts have

Recommended selection of fits (hole basis) (see also Appendix 4).

| Type of fit | | Shaft and tolerance | Hole and tolerance | | | |
|---|---|---|---|---|---|---|
| | | | H7 | H8 | H9 | H11 |
| Clearance | Slack ↑ / ↓ Close run | c 11 | | | | ▨ |
| | | d 10 | | | ▨ | |
| | | e 9 | | | ▨ | |
| | | f 7 | | ▨ | | |
| | | g 6 | ▨ | | | |
| | | h 6 | ▨ | | | |
| Transition | | k 6 | ▨ | | | |
| | | n 6 | ▨ | | | |
| Interference | | p 6 | ▨ | | | |
| | | s 6 | ▨ | | | |

From this the reader will be able to visualise the maximum and minimum metal conditions for the combinations in the table.

the nominal size as their upper limit and tend to give close running fits when associated with H holes. The j shafts have their limits disposed above and below the nominal and tend to give a fit between a clearance and an interference (transition fit). The k to z range lie above the nominal and give varying forms of interference fit. For general engineering practice a selection of shafts is recommended which, when associated with the H holes previously discussed, will provide for most of the fits required. The chief of these are set out in the table below together with the most suitable associated hole and the approximate fitting conditions. Details of the dimensional limits for these shafts are given in Appendix 3.

The diagram in Appendix 4 shows the hole and shaft tolerance relationships for the selection of shafts and holes in question.

*Example 1. To compare the former Newall 60 mm class 'B' hole-driving fit shaft combination with the ISO H8 hole—s7 shaft combination.*

(The Newall system of limits was in use up to about 1955/60)

Limits for 60 mm Newall B hole were   60·031

59·982 mm

,,     ,,     ,,     ,,   D shaft were 60·063

60·038 mm

Max. interference = largest shaft − smallest hole = 60·063 − 59·982

= 0·081 mm

Min.     ,,     = smallest ,, −largest   ,, = 60·038 − 60·031

= 0·007 mm

Limits for BS 60 mm H8 hole are 60·046

60·000 mm

,,     ,, ,,     ,,   s7 shaft ,,  60·083

60·053 mm

Max. interference = 60·083 − 60·000 = 0·083 mm

Min.     ,,     = 60·053 − 60·046 = 0·007 mm

Hence:

| | Interference (mm) | |
|---|---|---|
| | Max. | Min. |
| Newall B/D combination   .   . | 0·081 | 0·007 |
| BS H8/s7     ,,     .   . | 0·083 | 0·007 |

**Use of limits and tolerances**

ISO limits are not used by all engineers and the limits on a drawing may not necessarily belong to this system. We have given it as representing the best and most common example of a 'ready made' limit system in this country for general work. Engineering firms may prefer to create and use limits and fits which they consider more suitable for their particular work. In any case, no system gives any guidance on the determination of a tolerance for purposes outside the fit of one component in another. Limits and tolerances on such elements as angles, tapers, hole centres, contours, etc., must be determined from experience and other considerations. The adoption and careful adherence to a reliable system of limits and fits is an essential part of any modern system of interchangeable production. It ensures that components made in one factory may be assembled with mating parts made in another and, by eliminating the necessity for the preliminary trial of a fit, parts may be made and held in stock until they are required. The application of a system of limits and fits is not confined to circular holes and shafts but may be applied to any set of conditions where a particular type of fit is required. It will be seen, therefore, that it may be used for keys and keyways, square and flat fitting combinations and similar applications.

## Selective assembly

The principle we have just described is that of full interchangeability in which any component assembles with any other component. Often special cases of accuracy or uniformity arise which might not be satisfied by certain of the fits given under a fully interchangeable system and resort is made to a scheme of selective assembly.

If particular accuracy is required it is cheaper to work to a reasonable tolerance, and sort the articles into size groups during inspection, than to impose such close working limits that the conditions are achieved during manufacture. An interesting example of this is given in the mating of aluminium pistons in motor-car cylinder bores. On a bore of 63 mm the best skirt clearance for a certain type of piston is 0·12 mm on the diameter. If we assume that there is a tolerance of 0·02 mm on the bore diameter $\left(63\cdot00 \, {}^{+0\cdot02}_{+0\cdot00} \, \text{mm}\right)$ and the same on the skirt of the piston $\left(62\cdot88 \, {}^{+0\cdot02}_{+0\cdot00} \, \text{mm}\right)$, the smallest piston in the largest bore would be $63\cdot02 - 62\cdot88 = 0\cdot14$ mm clearance, and the largest piston in the smallest bore would be $63\cdot00 - 62\cdot90 = 0\cdot10$ mm clearance. By grading and marking the bores and the pistons as shown they may be selectively assembled to give the conditions required.

|              | A     | B     | C     |
|--------------|-------|-------|-------|
| Cylinder bore | 63·00 | 63·01 | 63·02 |
| Piston       | 62·88 | 62·89 | 62·90 |

## Gauging and measurement

We must now give some additional consideration to the practical aspect of checking work and for our purpose we had better assume the two general types into which it is divided: (*a*) Interchangeable production controlled by limits; (*b*) general engineering jobs, toolroom work, etc. Limits and tolerances will be found, of course, on drawings for small-quantity general production, but it is unlikely in such cases that the checking of work will be done by any other means than with simple gauges and standard measuring instruments. The usual method for controlling large-scale production is by a system of gauges which exercises a check on mating and other important surfaces, ensuring that they do not fall outside the limits laid down. For general work, toolroom jobs, and for production work on a small scale, it would not be economic to expend capital on expensive gauging equipment so that simple standard gauges and measuring equip-

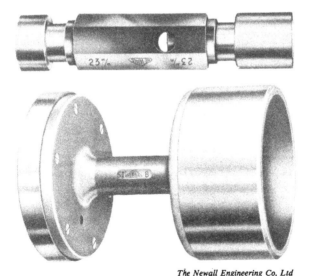

*The Newall Engineering Co, Ltd*

Fig. 6  Internal limit gauges

ment must be used. Indeed, since much of the toolroom work consists of making the gauges referred to above, together with other special and accurate manufacturing equipment, the toolmaker must rely on first principles for his accuracy. In production work, therefore, the tendency is for the fit of the gauge to be the important factor without much reference to any dimensional consideration and, as a matter of fact, it matters little whether a component is 25·00 mm or 25·25 mm provided it fits its mating part correctly, and that all spare parts have the same dimension. Toolmakers, and small-quantity production workers, who are making components into which the requirements of interchangeability do not enter, must tie themselves to fundamental methods, and the standard of length.

**Gauging**

As we have indicated, the use of gauges plays an essential part in the control of size and fit in any scheme of quantity production. A well-designed system of gauging enables different parts of the same assembly to be made in different factories, and even in different parts of the world, with the assurance that they will mate together accurately.

The function of a gauge is to check the level of the surface or surfaces that are being gauged and to show whether the metal lies within the zone permitted under the limits allowed on the drawing. In the workshop, gauging is necessary as the operation proceeds, and is carried out by the operator or by some other person responsible. Before the work passes to another operation, or into store, it is usual to subject it to a further check

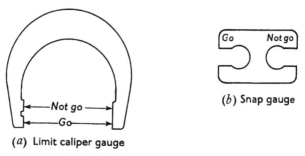

(a) Limit caliper gauge

(b) Snap gauge

Fig. 7 Limit gauges for shafts

by an independent inspector. Thus for the same component there may be two sets of similar gauges in use—the *working* gauge and the *inspection* gauge. The inspection gauge is often made so as to accept work slightly nearer the tolerance limit than the working gauge, so as to ensure that work which passes the working gauge will be accepted by the inspection gauge also. For example, if the 'Go' end of the working gauge were 25·045 mm, and corresponding inspection gauge 25·05 mm, for a shaft with upper size limit 25·05 mm, shafts which passed the working gauge would also pass inspection. Since gauges wear with use and may also become damaged, the working and inspection gauges must themselves be checked from time to time. Gauges for this purpose are called master, reference or check gauges. The master or reference gauge may not necessarily be similar to the gauges it must check; for example, to check a plug gauge a very accurate micrometer would suffice.

**Hole gauging.** The most usual gauge employed for holes is the limit plug gauge with 'Go' and 'Not Go' ends. Examples of these are shown at Fig. 6 where the type at (a) has solid ends, whilst that at (b), serving for larger sizes, has hollow ends to reduce weight. In each case the ends are detachable from the handle so that they may be renewed separately when worn, and to economise in cost, as the ends are made of a more expensive grade of steel. The 'Not Go' end is shorter than the other, as since it never enters the hole it need not have dimensions to minimise wear or prevent it jamming. The difference in length also helps to distinguish between the two ends.

**Shaft gauging.** Shafts and male components are usually checked by *caliper* or *snap gauges*, examples of which are shown at Fig. 7. The caliper gauge is used for the larger sizes and may be adjustable as shown at Fig. 8. This feature saves the expense of new gauges for small-quantity produc-

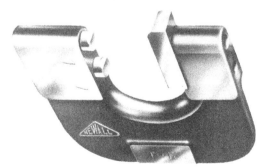

*The Newall Engineering Co, Ltd*

Fig. 8(a) Adjustable limit caliper gauge

tion, but as a precaution the gauge should always be set in the gauge room and sealed before issue.

**Wear of gauges.** However hard and wear-resistant the material of which a gauge is made, it will wear in the course of use, and if it is made to the lower (for a hole), or upper (shaft) limit when new, it will soon be accepting components outside the limits. This is often compensated for by allowing a percentage (often 10%) of the work tolerance for gauge wear. For example, if the 'Go' ends of the working and inspection snap gauges for a shaft of $25 \cdot 00 \, ^{+0 \cdot 05}_{+0 \cdot 00}$ mm were made 25·04 mm and 25·045 mm respectively, they could be allowed to wear to 25·045 mm and 25·05 mm respectively before it became necessary to replace them. We have already discussed the reason for the working limits being inside the inspection limits and, as the reader will appreciate, the effect of gauge wear is to reduce the actual working tolerance from that shown on the drawing. The precaution against wear is not so necessary for the 'Not Go' end of the gauge since, as it seldom passes over the work, its wear is very small.

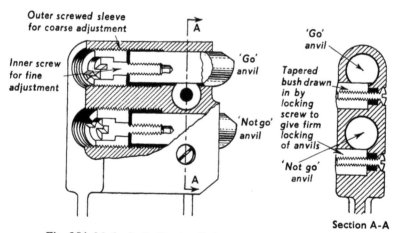

Fig. 8(b) Method of adjusting limit gauge shown at Fig. 8(a)

**Other limit gauges.** The snap gauge in a modified form may be used for gauging surfaces other than the diameters of cylindrical shafts, and at Fig. 9 are shown various applications of gauges of this type. Stepped limit gauges are useful in various capacities and some of these are shown at Fig. 10. Some limit gauges are of the type where conformation of the

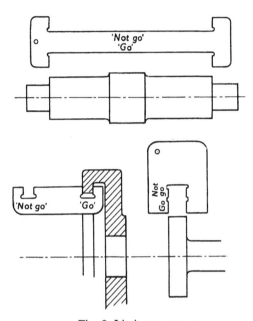

Fig. 9  Limit gap gauges

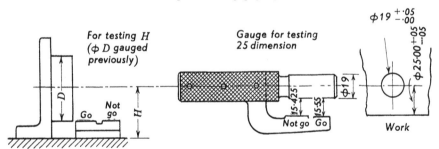

Fig. 10  Limit step gauges

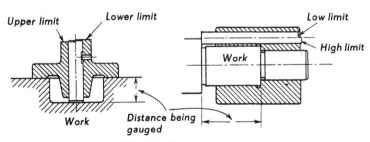

Fig. 11  Flush pin gauges

component to the limits is indicated by the surface in question, or by the end of a pin in contact with it standing at a level between those of two steps ground on the gauge. The difference in height between the steps is the tolerance and examination may be conducted by sight or feel (readers should try the 'feel' of a step of $\frac{1}{100}$ mm to realise their sensitivity). Examples of gauges using the end of a pin (called 'flush pin gauges') are shown at Fig. 11, whilst Fig. 12 shows the principle of indicating from the component surface itself.

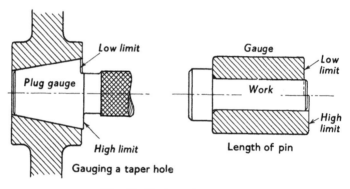

Fig. 12 Flush surface gauges.

**Indicating gauges.** It is a great advantage in gauging to be able to eliminate the personal element from the result obtained because, not only does this vary as between different people, but it also varies with the same individual, being affected by his health, temper, tiredness, etc. For this purpose many examples of mechanical, optical and electrical indicating gauges are in use. The pressure with which the gauging faces bear on the component is carefully controlled, is constant and independent of the operator. The size, or relation to the controlling + and − limits is recorded and transferred to an indicating dial by mechanical, optical, or electrical means. Thus the accuracy of the operation is made independent of the operator, and fatigue on his or her part reduced.

An example of one of these instruments is shown at Fig. 13. Movement of the plunger is indicated by a line on an illuminated circle which travels round the translucent scale. The magnification is ×1000, i.e. 5 mm on the scale corresponds to a plunger movement of 0·005 mm. The plunger is initially set with a master gauge to the size required and the indicating line set to zero. The tolerance indicators are then set to the permissible

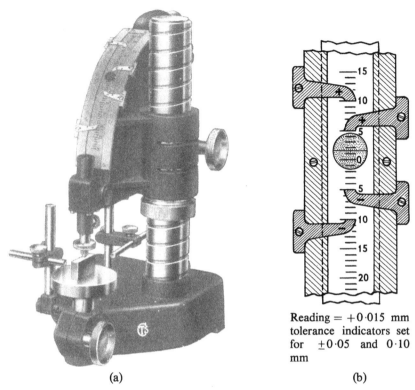

Reading $= +0.015$ mm
tolerance indicators set
for $\pm 0.05$ and $0.10$
mm

(a)                                        (b)

Fig. 13 Example of an inspection instrument which indicates the size condition
of the work

amounts above and below the zero position, so that when a component is
inserted its acceptance or rejection is observed at once.

**Profile and contour gauges.** To check the profile of parts some form of
profile gauge is necessary, and examples of such gauges are shown at Figs.

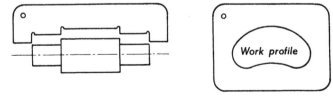

Work profile

Fig. 14                              Fig. 15

14 and 15. Fig. 14 shows a gauge for a stepped shaft, whilst Fig. 15 shows a receiver gauge for a component of special profile. The general point to be noticed about these gauges is that they either pass or reject the component without giving much information as to the excess or deficiency in metal on the surfaces being gauged. The gauge shown at Fig. 16 is for the groove turned in the bush shown, and incorporates in its design a slot for the purpose of setting the tool with which the groove is turned.

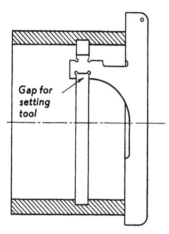

Gap for setting tool

Fig. 16  Gauge for groove in bush

**Special gauges.** The subject of gauges and gauging is a very wide one, and in the space at our disposal we have only been able to indicate its chief principles. To give the reader an idea of the appearance and function of a few special gauges we show a selection at Fig. 17. The gauges illustrated serve the following functions:

At (a) is shown a fixture incorporating the principles of the sine bar and suitable for checking taper plug gauges. The lower plug is a known distance from the base and the fixture may be tilted to one-half the taper angle by introducing a pile of slip gauges of the correct amount (see p. 57) underneath the other plug as shown. When set at the half-angle of taper the upper surface of the gauge should be parallel with the base and this may be checked with a dial indicator. Naturally, the taper surface must be true with the centre holes, and this may be verified beforehand by rotating it under the dial gauge.

The gauge at (b) is an alignment gauge for checking the alignment of a bore with a slideway which runs at an angle to it. At the same time the gauge checks the size of the slideway portion. Although the photograph

(a) Sine bar fixture for checking taper gauges

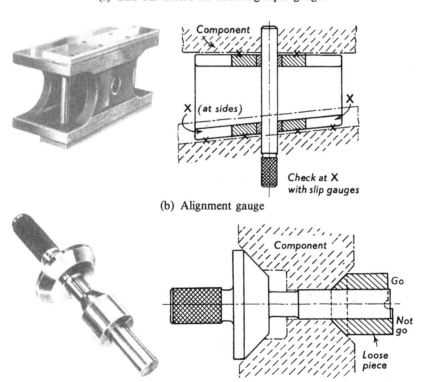

(b) Alignment gauge

The Pitter Gauge & Tool Co, Ltd

(c) Valve seat gauge for cylinder head

Fig. 17 Examples of special gauges (see text)

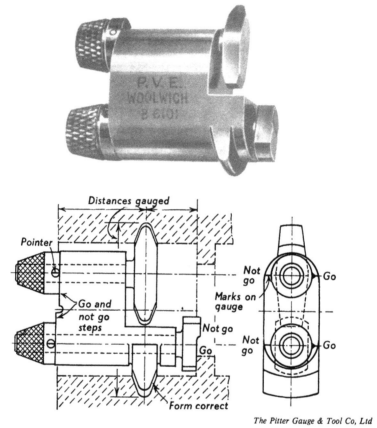

The Pitter Gauge & Tool Co, Ltd

(d) Gauge for groove and thrust face turned in a bore

Fig. 17 (*continued*)

makes the gauge appear parallel, and the hole square through it, there is really some taper and angle present, and a diagram showing the application of the gauge is shown alongside the photograph. After the gauge has been slid into position a test plug is inserted as shown and accuracy checked by means of 'Go' and 'Not Go' slip gauges between the surface of the job and the gauge at the points marked X. The vertical cross-bars on the gauge are merely for use as handles.

The gauge shown at (c) is for the valve seatings in a cylinder head. This gauges concentricity of seat with respect to bore as a blueing gauge, and checks the distance between the seat faces by means of high and low limit steps on the loose piece. The method of applying the gauge is shown in the diagram alongside, and it is necessary for the stem of the gauge to be a good fit in the bore, and the loose piece a good fit on the stem. Accuracy and concentricity of the valve faces are verified by applying a little engineer's blueing to the gauge and observing its distribution over the mating faces of the component after rotating the surfaces in contact.

Fig. 17(d) shows a gauge for checking the relative positions of a thrust face and groove turned in a bore. The diagram alongside shows the gauge in position, and it is manipulated as follows: With the end pointers midway between the 'Go' and 'Not Go' marks the gauge is inserted and allowed to position itself with the solid form piece on underside of the gauge body seating in the groove. The upper knob is then turned first to the 'Go', and at the same time observation made that the end of the work lies between the 'Go' and 'Not Go' steps on the end of the gauge body. The 'Go' and 'Not Go' of the lower gauging spindle are now tested by rotating it, and finally a check is made that the 'Not Go' property of the upper gauging element is satisfied.

**First operation gauges.** The machining of the first face on a casting or forging is often a very important operation because many of the subsequent operations depend on this face for their location. The amount of metal removed from this face influences not only other surfaces but also such things as the location of holes in the centre of their bosses, etc. Nothing looks

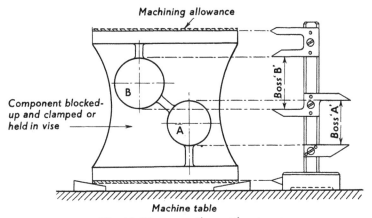

Fig. 18 First operation setting gauge

worse than a casting in which the holes are not central in the bosses provided for them. The use of a simple setting gauge will often help to equalise matters concerned with the machining on faces, relative position of bosses, etc., and avoid subsequent operations or bad work. Also, if a certain face has not sufficient metal on it to clean up, the fact is shown before time and expense have been incurred. Of course, the castings could be marked out, but this is a lengthy and expensive operation, to be avoided where possible. A diagram of a first operation setting gauge is shown at Fig. 18, being applied to a component during the process of wedging up and setting for the first operation of machining the top face. The function of the various indicators on the gauge will be obvious to the reader.

# 2 Measurement and precision work

The application of manufacture controlled by a system of gauges such as we have just discussed is mainly confined to production work where the machining and other processes have, as far as possible, been carefully subdivided into relatively simple operations which, with the assistance of special jigs and tools, can be performed by semi-skilled labour. It is true that not all production processes can be performed by any but a skilled operator, but in normal times the competitive urge of price reduction leads planning and workshop engineers to reduce, as far as possible, the skill necessary, so that lower paid labour may be applied.

There is still, however, a large volume of our work which cannot be treated in this way; the toolroom, in which the gauges, tools and appliances are made for the rest of the factory, the machine repair shop, the plant making machines and assemblies in small quantities to special order and so on. Except for the help of such standard gauges as they may find useful, e.g. plugs, slips, etc., these people must rely for their accuracy on fundamental methods of setting and measurement, and the reader will appreciate that the toolroom, which makes gauges and tools for the rest of the plant, must be the most accurate of all!

**Length standards—slip gauges***

It is essential that any shop in which finished engineering construction is undertaken should have some reference standards of length. The accuracy of such standards will depend, of course, on the limits imposed on the work going out; for example, in a foundry, where limits of accuracy are rarely closer than $\pm\frac{1}{2}$ mm, the rule can be the standard, whilst for toolroom work, where the tolerance may be as low as $\frac{1}{1000}$ mm, a much more accurate reference is necessary. To suit the general needs of most engineering workshops slip gauges and length bars are the most convenient and useful method of carrying such standards. Slip gauges, often called Johannsen gauges after their originator, are rectangular blocks of steel having a cross-section of about 32 mm by 9 mm which, before being

* BS 4311: 1968.

finished to size, are hardened and carefully matured so that they are independent of any subsequent variation in shape or size. (The longer gauges in a set, as well as the length bars, are only hardened locally at their ends.) After being hardened the blocks are carefully finished on their measuring faces to such a fine degree of finish, flatness and accuracy that any two such faces when perfectly clean may be 'wrung' together. This is accomplished by pressing the faces into contact and then imparting a small twisting or sliding motion whilst maintaining the contact pressure. When two gauges are wrung together they adhere so that considerable force is necessary to separate them, and the overall dimension of a pile made of two or more blocks so joined is exactly the sum of the constituent gauges. It is on this property of wringing units together for building up combinations that the success of the system depends, since by combining gauges selected from a suitably arranged combination almost any dimension may be built up.

Slip gauges are made in five grades of accuracy: Grade I, Grade II, Grade 0, Grade 00 and Calibration Grade. The grade most commonly used in the production of components, tools and gauges is Grade I but for rougher checking Grade II may be suitable. Grades 00 and Calibration should not be used for general work, but to check other gauges and standards. Slip gauges are finished on their measuring faces to a very high degree of accuracy and the permissible departures from absolute accuracy for the first three grades is given in the table below:

Slip gauges. Maximum permissible errors Unit $= \frac{1}{100\,000}$ mm

| Size of Gauge (mm) | | Grade II | | | Grade I | | | Grade 0 | | |
|---|---|---|---|---|---|---|---|---|---|---|
| Over | Up to and incl. | F[1] | P[2] | Gauge length | F | P | Gauge length | F | P | Gauge length |
| — | 20 | 25 | 35 | +50 −25 | 15 | 20 | +20 −15 | 10 | 10 | ±10 |
| 20 | 60 | 25 | 35 | +80 −50 | 15 | 20 | +30 −20 | 10 | 10 | ±15 |
| 60 | 80 | 25 | 35 | +120 −75 | 15 | 25 | +50 −25 | 10 | 15 | ±20 |
| 80 | 100 | 25 | 35 | +140 −100 | 15 | 25 | +60 −30 | 10 | 15 | ±25 |

[1] F = Flatness      [2] P = Parallelism

*The Coventry Gauge & Tool Co, Ltd*

Fig. 19 A set of slip gauges (107 pieces) (lid of box not shown)

Slip gauges are supplied in sets, the size of which varies from a set of about 112 pieces down to one containing 32 pieces. A set of gauges containing 107 pieces is shown at Fig. 19.

The grade and the size of set desirable depend on the class of work for which they are required, the number of combinations that should be available, and the price one is prepared to pay. An M88 Grade I set should cope with all the requirements of the average workshop and this contains the following gauges:

Gauges contained in an M88 set of slips
(sizes in millimetres)

| Size or range | Increment | No. of pieces |
|---|---|---|
| 1·0005 | — | 1 |
| 1·001–1·009 | 0·001 | 9 |
| 1·01–1·49 | 0·01 | 49 |
| 0·5–9·5 | 0·5 | 19 |
| 10–100 | 10 | 10 |
| | Total | 88 |

**Building up sizes with slip gauges**

Working with the above set it is possible to assemble a combination to give any size in millimetres to the third place of decimals. The 1·0005 slip also enables digit '5' in the fourth place of decimals to be included, but for

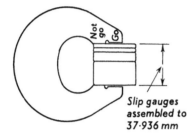

Slip gauges assembled to 37·936 mm

Fig. 20 Testing gap gauge with slips

ordinary purposes this latter is rather unrealistic. As a general rule, the least number of gauges to make up the combination should be used.

*Example 1. To assemble 72·225 mm from the above set.*

$$\begin{array}{ll} \text{1st slip} = & 1·005 \\ \text{2nd slip} = & 1·22 \\ \text{3rd slip} = & \underline{70·00} \\ & \overline{72·225} \end{array}$$

*Example 2. Combinations necessary to check the 'Go' and 'Not Go' dimensions of a limit gap gauge* $38\,{}^{-0·025}_{-0·064}$ *mm.*

(This is the size of a 38 mm BS f8 shaft)

| 'Go' Anvils of Gauge | 'Not Go' Anvils of Gauge |
|---|---|
| = 38 − 0·025 mm | = 38 − 0·064 mm |
| = 37·975 mm | = 37·936 mm |
| 1st slip  =  1·005 | 1st slip  =  1·006 |
| 2nd slip =  1·47 | 2nd slip =  1·43 |
| 3rd slip  =  5·50 | 3rd slip  =  5·50 |
| 4th slip  = 30·00 | 4th slip  = 30·00 |
| 37·975 mm | 37·936 mm |

A sketch of the slips inserted to test the gauge is shown at Fig. 20.

**English working**

For checking English sizes the metric equivalent can be calculated and a suitable combination of gauges assembled. For regular English working, however, it is advisable to obtain a set of gauges in inch units.

## Length bars

Slip gauges are not made in lengths above about 100 mm, and when greater lengths are required than can be built up with these gauges, length bars are available. These are circular in section, about 22 mm diameter, with their ends hardened and finished to an accuracy comparable with that of slip gauges. The BS recommendation for a set of workshop length bars consists of 11 pieces made up of 8 bars from 25 mm to 200 mm in steps of 25 mm, and 3 bars of lengths 375 mm, 575 mm and 775 mm. These bars are held end to end, when assembled into lengths, by means of studs screwed into tapped holes in their end faces.

**Care of slip gauges.** A set of slip gauges is such a fine piece of construction, and serves such a vital purpose, that every care should be taken to prolong its life and accuracy. When not in use the blocks should be kept in their case and when in use they should be in an atmosphere free from dust, those not in use being in the closed case. Before being wrung together their faces should be wiped with a clean chamois leather or linen cloth and the measuring face should not be fingered. If a slightest sign of roughness or scratching be felt during wringing stop immediately and examine the gauge faces for burrs or scratches. If before use the gauges have been handled for some time they should be allowed to settle down to the prevailing temperature of the room before a measurement is taken. When a pack of gauges is in use the outer gauging faces of the two end ones are subjected to most of the wear and abrasion. To protect these a pair of *protection slips* may be used. These are often included as an extra in sets of gauges, or a pair of hard, tungsten carbide slips in a separate case may be obtained.

After use the gauges should never be left wrung together. Slide the gauges apart, do not break the wring. Before returning the blocks to their case any finger-marks should be wiped off, and if they are not likely to be used again for some time a thin smear of grease may be applied, but care should be taken to use a reputable grease which is free from acid. Eventually the surfaces of slip gauges will wear and may suffer slight changes due to ageing in the steel. When there is any doubt of their accuracy they should be sent to the National Physical Laboratory who, for a nominal charge, will check them and issue a certificate showing their variations from the sizes marked on them.

### Size transference; comparators

As discussed later, the slip gauges, with a few simple accessories, can be used for obtaining a direct measurement, and may also be used alone for various examples of checking. Often, however, an instrument is needed to compare the dimension of some article to be checked with the length of a combination of slips assembled to the required dimension. Such an

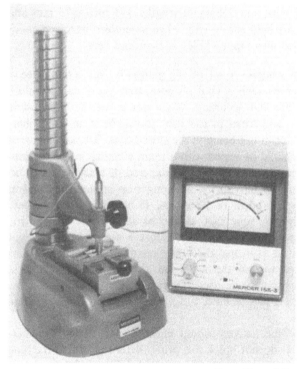

*Thomas Mercer, Ltd*

Fig. 21  Comparator with electronic gauging probe

instrument is called a *comparator* (comparer), because it does not incorporate any basic length standard such as a micrometer or vernier, but merely compares a given length with some dimension to which it has been set. Fig. 21 shows that the essential details of the construction of a comparator are a base and measuring table above which, carried on a standard, is a sensitive gauging head. The instrument shown carries an electronic probe as the gauging element and signals from this are fed into the gauge unit shown alongside. Five ranges of magnification may be selected by the control knob on the unit. These give magnifications of 320, 1050, 3200, 10 500 and 32 000 corresponding to full scale readings of $\pm 0 \cdot 150$ mm, $\pm 0 \cdot 050$ mm, $\pm 0 \cdot 015$ mm, $\pm 0 \cdot 005$ mm and $\pm 0 \cdot 0015$ mm. The RH knob on the unit is for zeroing the pointer before use.

The platen on the instrument shown is for the comparison of slip gauges but interchangeable platens for other forms of work are available.

When using a comparator, the dimension of the article to be measured is built up by wringing together a suitable combination of slip gauges, and with these placed under the measuring plunger the comparator is set to zero. The article to be measured is now substituted for the gauges and its similarity to or variation from their length is indicated on the dial.

The chief advantage of the comparator over some alternative methods of comparing a dimension is the reliability of the result. This is due to the inherent accuracy of the instrument and because the pressure used for measurement (between the plunger and the work) is carefully controlled and is the same for both the gauges and the work being compared. We could measure, say, a plug gauge by setting a micrometer to a suitable combination of slip gauges and then testing the plug with the micrometer. But our accuracy would depend on the uniformity with which we could 'feel' the gauges when setting and the plug when testing, and if a variation in size existed, on the capcity of the micrometer to indicate the variation accurately. By using a micrometer with a vernier attachment, reading to $\frac{2}{1000}$ mm (see p. 46), there is some assurance of accuracy in reading a variation, but the problem of obtaining uniformity in measuring pressure ('feel') remains, particularly in a case such as this where flat contact is used between the micrometer anvils and the slip gauges, and line contact between

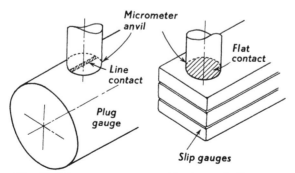

Fig. 22 Micrometer contact with plug and slip gauges

the micrometer and plug gauge (Fig. 22). It is advisable, therefore, when possible and when accuracy is important, to use some form of comparator for checking precision work.

### Roller gauges and test bars

In the average shop there are many testing and measuring jobs for which the extreme accuracy of slip gauges is unwarranted, and to use these gauges for any but essential purposes is detrimental, leading to their rapid wear. It is bad practice to allow such precise and expensive tools to degenerate into workshop hacks and such a state of affairs quickly develops unless alternative means of measurement are provided. Much useful work to very fine accuracy can be done with the help of balls and rollers supplied by ball bearing manufacturers. The firms no longer put up special sets of these but their standard products are very accurate and with the help of a good comparator it is not difficult to collect together a set coming within an accuracy of, say, 0·0025 mm. A useful set (duplicated) might consist each of balls and rollers rising from 4 mm×2 mm to 32 mm. When they have been collected a simple wooden case should be made to hold and protect them.

Test bars are necessary for measuring the centre distances of holes. The bars should be from 75 mm to 100 mm long on the working portion, with a short reduced end provided with a flat for affixing a small lathe carrier to serve as a handle. The bars may be of case-hardened mild steel and the

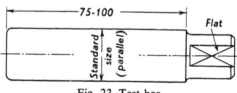

Fig. 23 Test bar

diameter should be ground to dead size and parallel (Fig. 23). The range of sizes necessary will depend of course on the range of hole sizes likely to occur, but a useful general purpose set might consist of two each of the following sizes: 6 mm by 1 mm to 25 mm, then by 5 mm to 50 mm diameter (i.e. 25 plugs duplicated). They should be stored in a wooden tray or box with a separate compartment for each one.

## Measurement auxiliaries and instruments

Before we discuss some examples of inspection and measurement we must mention a few instruments we have not previously covered. It is important in this work that we should possess the widest knowledge of the possibilities of every available piece of equipment, since there is often a choice of alternative methods for carrying out any given inspection job. When such a choice faces the reader he should, as a rule, adopt the simplest and most direct method provided it offers him unquestioned reliability in accuracy to the degree required. Some methods of working because they offer novel and rather academic features, often appeal to those who aspire to an academic approach to their work, but the reader should always bear in mind that, in the workshop, sound practice is the most important requirement.

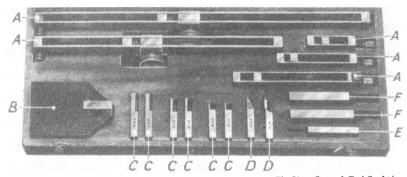

The Pitter Gauge & Tool Co, Ltd

Fig. 24. Slip gauge accessories.

*A*, 25, 60, 100, 200 and 300 mm slip gauge holder; *B*, Height gauge base; *C*, Jaws for internal and external measuring; *D*, Scriber and centre point; *E*, Straight edge; *F*, 25 mm sectional external jaws.

**Slip gauge accessories.** To enable the most effective use to be made of slip gauges various accessories may be obtained, the use of which greatly extends their scope whilst at the same time retaining the accuracy associated with them. A representative set of these is shown at Fig. 24, from which it will be seen that the chief units are holders, jaws, scriber point, centre point and base.

*Measuring jaws.* By assembling a pile of slips in a holder with external jaws at each end the combination may be used for an accurate straddle

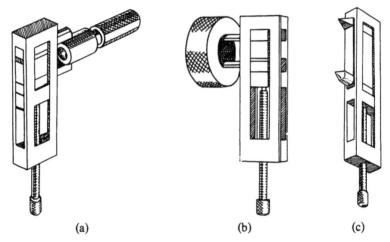

(a)             (b)             (c)

Fig. 25 Uses of slip gauge accessories

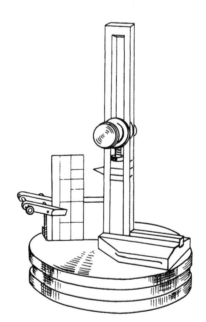

The Pitter Gauge & Tool Co, Ltd

Fig. 26 Holder assembled with base for height gauge work

measurement as suggested at Fig. 25(a). The internal jaws enable an internal size to be checked as at (b).

*Scribing and centre points.* These have a variety of uses. The face of the centre point which bears on the slip gauge is exactly on the point centre so that a pair of points at the ends of an assembly of gauges becomes an accurate point spacer, whilst a point at one end and a scriber at the other is equivalent to a pair of trammels or dividers for scribing accurate circles (Fig. 25(c)).

*Base.* Used with one of the holders and the scribing point or I-section jaws, the base can be assembled to form a height gauge which may be used for scribing lines or checking a surface according to whether the point or jaw is placed on the top of the slips. The use of this for scribing is shown at Fig. 26.

## The vernier height gauge

The vernier height gauge incorporates its own vernier measuring scale and is practically equivalent to a large vernier caliper with its fixed jaw set rigidly into a base. The reading of the vernier is that of the height of the upper surface of the sliding jaw above the base. The jaw itself may be used for testing underneath surfaces and test plugs, and when the finger is clamped on this may be used for feeling over surfaces and its sharp end may be used to scribe lines. When testing a plug the most reliable feel may be obtained by discarding the finger and working to the *underside* of the plug, but on no account should any force be used. During such a trial the main jaw should be moved with the adjusting screw until, when the jaw is slid under the plug, contact can just be felt. Whilst this adjustment is taking place the clamping screw of the main jaw should not be loose but lightly nipped to such a tightness that the adjusting screw can still move the jaw, and when the feel is to a nicety this screw should be tightened, and another trial of the feel made before taking the reading. A similar procedure should be followed when using the finger to take readings on upper surfaces. A vernier height gauge and some of its uses are shown at Fig. 27. On the Chesterman gauge shown, adjustment of the jaw is made by the thumbscrew on the base, which operates a screw passing up the inside of the column and engaging with a nut on the slide.

The height gauge is a very useful instrument in any shop, but it should not be used on work for whose accuracy the scribing block would be adequate. It is not a robust instrument, should be treated with care, and without being tested should not be relied upon too much for the accuracy

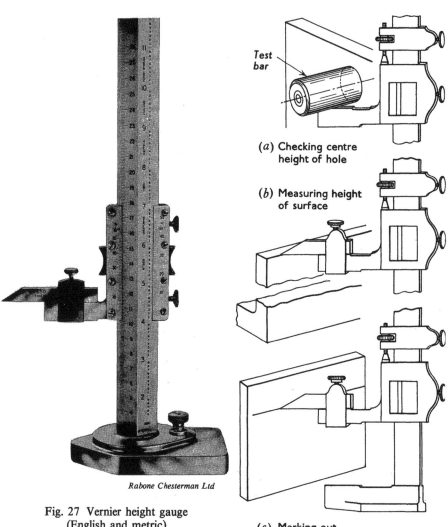

Test bar

(a) Checking centre
height of hole

(b) Measuring height
of surface

Rabone Chesterman Ltd

Fig. 27  Vernier height gauge
(English and metric)

(c) Marking out

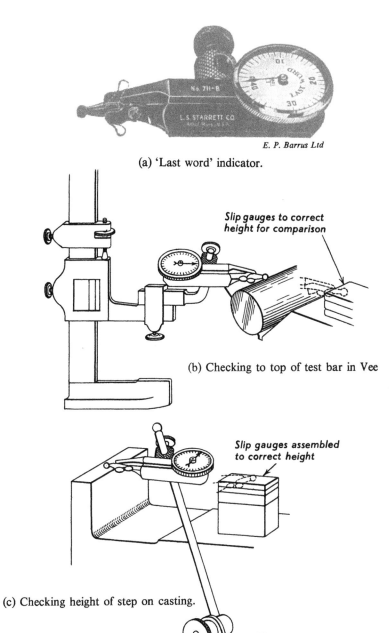

(a) 'Last word' indicator.

*E. P. Barrus Ltd*

*Slip gauges to correct height for comparison*

(b) Checking to top of test bar in Vee

*Slip gauges assembled to correct height*

(c) Checking height of step on casting.

Fig. 28 Indicator applied to serve as portable comparator

of its readings of absolute heights from its base. The slightest straining of the base or jaw unit by rough usage or dropping is enough to destroy its accuracy in height indication from the base. Even if the instrument *is* at fault in this respect, however, it will still be accurate for measuring *differences in height*, a function for which it is used probably more than for measuring absolute heights from its base. The method of reading the vernier height gauge is similar to that for the vernier calipers which we discussed on page 220 in Part I. (When using the height gauge the reader should cultivate the habit of writing down readings and adding and subtracting on paper. This helps to avoid mistakes caused by mental working.)

*The height gauge applied as a comparator.* For some purposes the height gauge may be usefully employed as a kind of portable comparator for the purpose of comparing some surface with the height of a gauge, or with a pile of slip gauges. The variable and doubtful property of the feel of the finger may be eliminated and increased accuracy ensured if the finger is replaced by a small dial indicator with a ball-ended lever attachment which is set on to the gauges and then transferred to the surface to be measured. Equality in height or any difference existing will be indicated by the clock when it is transferred to the surface. Actually, of course, in this application, the height gauge is merely serving as the carrier of a dimension, its own measuring scale playing no part in the process. A diagram of the Starrett 'Last Word' Indicator and its applications is shown at Fig. 28.

**Use of the height gauge for marking out**

The finger of the height gauge is ground to a sharp edge, and as the height of this coincides with the reading of the vernier scale, lines scribed with it may be relied upon to close limits of accuracy. This feature of its construction enables it to be used for marking out such components as templates, gauges, press tool dies and other work in which a fine degree of accuracy is essential. The following examples will illustrate marking out of this nature.

(a) *To mark out the template shown at Fig. 29(a)*

This would probably be made from gauge steel about 3 mm thick, which would come for marking out with both faces smooth (probably bright ground), and in a piece slightly larger than the overall sizes of the template.

First coat one face of the material with copper sulphate solution so that when it is dry a thin layer of copper is deposited which will cause the scribed lines to show up. Secure the work to an angle plate, using a plate

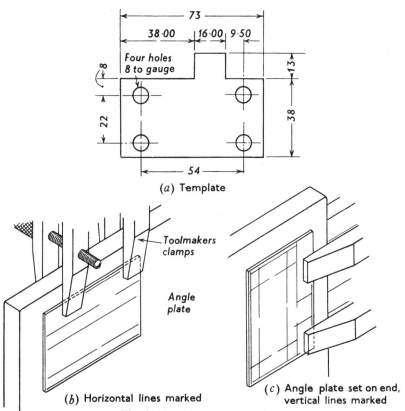

(a) Template

(b) Horizontal lines marked

(c) Angle plate set on end, vertical lines marked

Fig. 29  Marking out a template

with one end machined square with the working faces so that it may be turned over. Small bolts and clamps or toolmakers' clamps may be used to hold the work, and it should be set with the long edges approximately horizontal.

If the material is more than 51 mm wide set the finger of the height gauge about half the excess above the lower edge and scribe a line across the plate. (Should the blank be just 51 mm wide, use the gauge to set one of the long edges exactly parallel with the base of the surface plate, and then set the finger level with the bottom edge.) Raise the height gauge 38 mm and scribe the line for the upper edge. Write down this reading and then raise a further 13 mm for scribing the top of the tongue portion. Subtract 8·00 mm from the reading written down for the height of the top edge, set the gauge to this and scribe two lines for the upper hole centres.

Subtract a further 22·00 mm and scribe the lower hole centre lines (Fig. 29(b)).

Turn the angle plate on to its end and set the height gauge above the lower end of the material by about one-half of the excess of the material over 73 mm. Scribe the lower end and write down the reading of the gauge. Add 38·00 mm to the reading, raise the gauge to this height and scribe the lower side of the tongue; raise a further 16·00 mm and finish the tongue, raise a further 9·50 mm and scribe the centres of the upper holes. Subtract 54·00 from the last setting, lower the gauge and scribe the centre lines of the lower holes. To finish the job the gauge may now be set 73 mm above the reading for the bottom edge, and a line scribed for the upper end (Fig. 29(c)). The work may now be removed from the angle plate, the outline dot punched and the hole circles scribed.

*(b) To mark out the die profile shown at Fig. 30(a)*

Brush the face of the die with copper sulphate and fasten to an angle plate as before.

Set the lower edge of the die parallel with the surface plate, using the height gauge underneath it, then raise the gauge by one-half the width of the die, fit the finger and scribe the centre line. Raise, and lower the finger by 13·50 mm from the centre line setting and scribe the horizontal centre lines of the holes. Turn the angle plate on to its end and mark the vertical centre line from the holes 22 mm from the lower edge of the die (Fig. 30(b)). Now, with a vernier protractor or sine bar, set the die to an angle of 45°, setting from the edge which was levelled up in the first instance. Using a magnifying glass set the finger of the height gauge to the centre of the lower hole. Reduce the reading by 7·25 mm and scribe the inner line of the angle, reduce by a further 13 mm and scribe the outer line, increase by 38 mm and scribe the end line of the other side (Fig. 30(c)).

Turn the angle plate on its end, pick up the centre of the other hole and mark the remaining lines as before (Fig. 30(d)).

This completes the marking out and the die may be removed from the angle plate to have the circles scribed, and profile of marking out dot punched.

**The vernier micrometer**

By applying a vernier scale to a micrometer as an addition to its usual barrel and thimble markings, the rotation of the barrel between any two of its graduations may be divided into 5 parts and thus readings to $\frac{2}{1000}$ mm

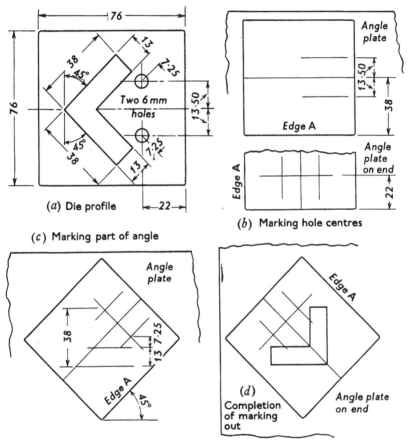

(a) Die profile

(b) Marking hole centres

(c) Marking part of angle

(d) Completion of marking out

Fig. 30 Marking out a die-profile

made possible. Diagrams of this are shown at Fig. 31 on which the reading is indicated, and the drawing shown should convey the principle of the scale without further explanation. The reader should not imagine that he could take one of these micrometers and proceed to carry out reliable measurements to $\frac{2}{1000}$ mm. A five-hundredth is a minute amount and even if the instrument were perfect for the whole travel of its motion, had not been dropped or maltreated and was at the same temperature as the work, to measure with it to such a degree is very fine work indeed and demands a performance beyond the capacity of most of us. Indeed, we are sure that many experienced readers will share our opinion that when accuracies of

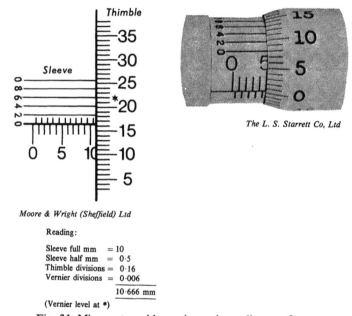

*Moore & Wright (Sheffield) Ltd*

*The L. S. Starrett Co, Ltd*

Reading:

| Sleeve full mm | = 10 |
|---|---|
| Sleeve half mm | = 0·5 |
| Thimble divisions | = 0·16 |
| Vernier divisions | = 0·006 |
| | 10·666 mm |

(Vernier level at *)

Fig. 31 Micrometer with vernier scale reading to $\frac{2}{1000}$ mm

this order are necessary the micrometer is not the most suitable instrument to use.

## The depth gauge

For measuring the depth of a step or recess a depth gauge is useful and such gauges may be obtained operating either on the micrometer or on the

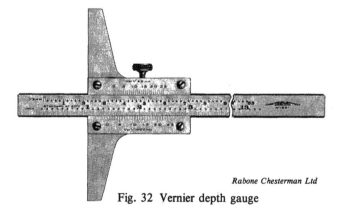

*Rabone Chesterman Ltd*

Fig. 32 Vernier depth gauge

vernier principle. A vernier depth gauge is shown at Fig. 32, and the micro-meter instrument is similar in construction except that the measuring plunger is moved by a micrometer head. For deep readings the micrometer instrument must have an arrangement for setting out the plunger a certain specified amount to start with, as the travel of the operating screw will only be 13 mm or 25 mm.

## Examples of inspection

By reason of their variety, and the fundamental nature of their presentation, the less straightforward testing and measuring jobs of a general shop form one of the most interesting parts of its technique. If he is to achieve any measure of success in this work, or for that matter in any of the higher branches of his profession as a workshop engineer, the reader must develop a scientific approach and the capacity to stick at a job to its successful completion. A man who cannot apply his whole mind to the work in hand, who starts one job and before it is finished drops it to commence some-thing else, who is too impatient to give the protracted attention often neces-sary to carry a job to its finish, and so on, will never make a successful workshop engineer and if, by some accident, such a man reaches a respon-sible position in the profession he will never possess that measure of stability so necessary in the minds of the leaders in industry.

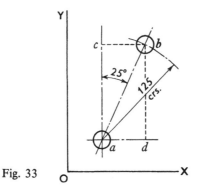

Fig. 33

To revert to our subject, there are often more ways than one of carrying out a check, and if this happens the method which promotes the greatest accuracy in the result should always be used. Let us consider, for example, the simple case of checking the angular position of hole *b*, relative to *a*, in Fig. 33. This can be done by checking perpendicular to face OX and measuring distance *bd*, or by working from OY and checking *bc*. The trigonometry of the problem shows us that *bc* should be 125 sin 25° = 52·83 mm and *bd* = 125 cos 25° = 113·29 mm. Now let there be an error of −15′ in the angle, i.e. let the angle be 24° 45′ instead of 25°. The length *bc* will now become

$$125 \sin 24° 45' = 52\cdot33 \text{ mm}$$

whilst *bd* will become

$$125 \cos 24° 45' = 113\cdot52 \text{ mm}$$

Due to this slight error in the angle, the changes in *bc* and *bd* have been:

$$\text{in } bc: \ 52\cdot83 - 52\cdot33 = 0\cdot50 \text{ mm}$$
$$\text{in } bd: \ 113\cdot52 - 113\cdot29 = 0\cdot23 \text{ mm}$$

Thus *bc* is more than twice as sensitive as *bd* for indicating variations in the given angle, and when locating *b* for boring, or when checking its position, we should work in the direction of *bc* for the promotion of maximum accuracy, and for discovering any error most easily.

For our first example we might consider the bracket shown at Fig. 34 in which two holes, spaced as at Fig. 33, are incorporated. The steps in checking this might conveniently be carried out as follows:

1. Try the base and vertical machined side for flatness against the sur-

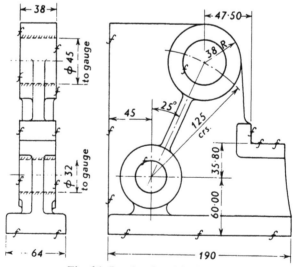

Fig. 34 Bracket (machined at *f*)

face plate and with a try-square verify that the left-hand vertical face and the sides of the bosses are at right angles to the base.

2. Fit a 32 mm test bar into the lower hole, stand the bracket on its base and check the 60·00 mm dimension. This can be done by measuring between the plug and surface plate with slip gauges or height gauge. The measurement between the bottom of the plug and the surface plate should be $60·00 - \dfrac{32}{2} = 44·00$ mm. Allow the test bar to project both sides and test each end for a check of parallelism.

3. Check the height of the machined step from the base. This can be done either with a 75–100 mm micrometer or with a height gauge from the surface plate. Check also for parallelism. If the 60·00 mm dimension is correct the step height should be $60 + 35·80 = 95·80$ mm.

4. Fit a 45 mm test bar to the top hole and check the 125 mm dimension by measuring over or between the plugs. Slip gauges may be used for inside checking or a micrometer, or vernier if the overall dimension is checked. (Check for parallelism by trying at both sides of bracket.) If the 125 mm is correct the inside measurement between plugs should be $125 - \dfrac{45}{2} - \dfrac{32}{2} = 86·50$ mm (Fig. 35(a)).

5. Turn the bracket on to its vertical edge and check the 45·00 mm dimension for the bottom hole. Now check the 25° angle by measuring the vertical distance between the centres of the holes, which from Fig. 33 should be 52·83 mm. The vertical distance between the bottoms of the plugs should be $52·83 - 22·5 + 16 = 46·33$ mm. These checks may be made either with slips from the surface plate, or height gauge (Fig. 35(b)).

6. Check the 47·50 mm dimension to the edge of the step. This can be done with a height gauge, working from the plug in the 45 mm hole or by slips between the plug and something (e.g. the blade of a square) held against the face (Fig. 35(c)). (Test both ends for parallelism.)

7. Check the boss faces for parallelism with a micrometer (their dimension is not important), and with a straightedge verify on each side that their faces are in the same plane.

This completes the checking of the important details and the remainder of the dimensions, metal thickness, symmetry of bosses, etc., may be checked with a rule.

The shaft shown at Fig. 36(a) involves methods of examination commonly encountered with many components. It is ground on all diameters and the position of the keyway relative to the squared portion is important.

As it will be ground on its own centres, the roundness and concentricity of the diameters may be accepted almost without question, so that the first step would be to check the diameters for size and parallelism. This could be done by means of micrometers. Next, the lengths could be verified, a rule being sufficient for all but the $\frac{60 \cdot 00}{59 \cdot 90}$ mm step. For this there are various methods available: (*a*) The shaft could be stood upright by resting it on its left-hand end, or placing the end diameter in a loose sleeve so that the shaft is supported on face A. With the shaft in this position the step can be

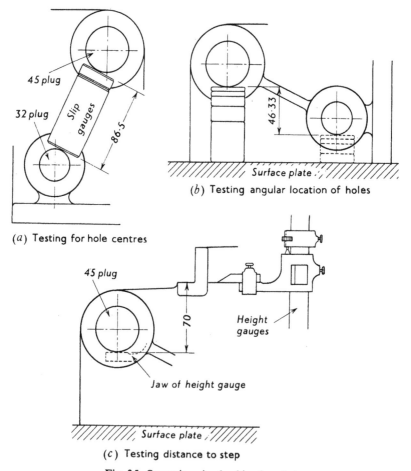

(*a*) Testing for hole centres

(*b*) Testing angular location of holes

(*c*) Testing distance to step

Fig. 35  Operations in checking bracket

checked with a height gauge. (*b*) By measuring distances BD and BC with a vernier and subtracting, the distance CD is obtained. (*c*) A depth gauge could be used from D to C. At the time of checking the length steps the flatness and quality of the machining on the shoulders should be examined.

The squared portion of the shaft may now be checked for size, shape, axial alignment and concentricity, and for this a pair of centres (Part I, Fig. 144) or the use of a lathe bed and centres is necessary. First check the square for size and parallelism with a micrometer, and then verify the squareness of its faces with a try-square. To test for concentricity and alignment place portion with the jaw of the height gauge, adjusting the shaft and height gauge until the face being tested is horizontal in the cross direction, i.e. until the feel is the same at front and back. Lock the height gauge and test the flat for axial alignment by feeling for its parallelism longitudinally. Now

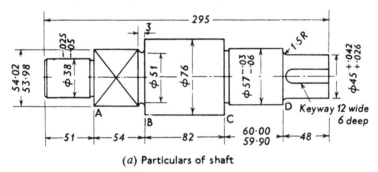

(*a*) Particulars of shaft

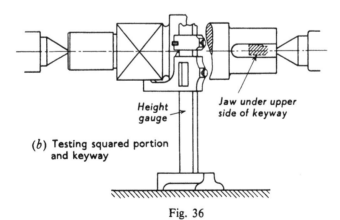

Height gauge

Jaw under upper side of keyway

(*b*) Testing squared portion and keyway

Fig. 36

rotate each of the other flats into position and test them in turn. An alternative method of making this test would have been to work on the upper side of the square with a dial indicator.

We now have to check the keyway for size, and for its relative position with the squared portion of the shaft. Its size may be measured with a plug or slip gauge, and its depth by measuring across from its bottom face to the opposite side of the 45 mm diameter which distance should be $45 - 6 = 39$ mm. To test the radial centre line of the keyway for being perpendicular to the side of the square, place the shaft between centres again and set the height gauge underneath the flat which is on the lower side, when the keyway is horizontal, and adjust the shaft until the feel of the gauge is the same all over the flat. Read the height gauge and add to its reading half the square (27 mm), and half the width of the keyway (6 mm). When set to this new height the finger should just be level with the upper side of the keyway (Fig. 36(b)).

This completes the examination of the shaft except for the 1·5 mm radius

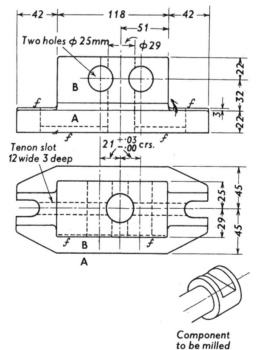

Component
to be milled

Fig. 37 Milling fixture casting (not fully dimensioned)

at the end of the 45 mm portion which may be checked with a rule or radius gauge. Before passing the shaft it should be visually examined to ensure that the sharp edges at the extremities of its diameters have been taken off, as these can be dangerous to persons handling the shaft.

The machined casting shown at Fig. 37 is the body of a fixture to be used for milling the flat on the head of a pin as shown in the inset. Two pins are milled at one setting and they are held by their shanks fitting in the two 25 mm holes, the underside of their heads locating against the vertical machined face of the holes, and a pad bolt which fits in the centre vertical hole clamps and prevents them from turning. The fixture is lined up on the table of the milling machine by key pieces, which fit in the centre tee slot and locate in the 12 mm tenon slot running through the base of the fixture, so that the arrangements for positioning the work must be accurately aligned with this slot. To assist in obtaining alignments with a slot of this type (commonly used on milling fixtures) during the machining of the casting it is helpful if, at the time that the slot is machined in the base, one of the edges of the base (e.g. edge A) can be cleaned up. If this is done at the same setting there will be no doubt as to the parallelism of the slot and the machined edge and the latter is much more convenient than the slot for use as a location during subsequent settings up. We will assume that edge A has been cleaned up but will not trust to its being parallel with the slot until we have verified the fact with a vernier. In testing the machining of this casting we could proceed according to the following sequence:

(a) Check the base for flatness, the width of the tenon slot with a plug or slip gauge, its depth with a rule, and with a vernier confirm that the side of the slot is parallel with the machined edge of the base. With a rule check the holding down bolt slots and verify that they are central with the 12 mm tenon slot.

(b) The front machined face (B) of the holes must now be checked for its squareness with the base and its longitudinal alignment with the tenon slot. Its squareness can be checked by clamping the face to a parallel strip as shown at Fig. 38(a) and placing a try-square against the base. Its alignment with the tenon slot may be checked at the same setting by verifying the parallelism of the tenon slot and the surface plate, using a height gauge. An alternative method of checking this would be to use a depth gauge from face A, making a trial at each end.

(c) The centre lines of the two 25 mm holes which carry the components should be parallel with the base, perpendicular with the tenon slot and at the same height. Their actual height is relatively unimportant and their

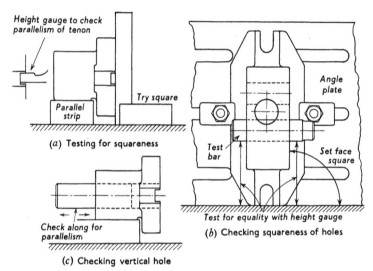

(a) Testing for squareness

(b) Checking squareness of holes

(c) Checking vertical hole

Fig. 38  Operations for checking casting

centre distance is only of sufficient importance to ensure that the pad bolt functions satisfactorily.

Check the holes for size with a 25 mm plug gauge and fit test bars in each, allowing the bars to protrude from both sides. With a height gauge check for their uniformity in height and for being parallel with the base. With slip gauges between the bars at each end, check the holes for parallelism in vertical planes. To check these holes for squareness with the tenon slot the casting may be clamped to an angle plate with face A square and the test bar checked with a height gauge (Fig. 38(b)). The centre distance of the holes may be checked by noting the distance between the test bars when checking for parallelism, and adding 25 mm to this amount.

(d) Check the pad bolt hole for size with a plug gauge and verify that it is centrally placed between the other two holes. This can be done at the setting shown by Fig. 38(b) by using the height gauge underneath a test bar fitted to the hole. The hole should be central within one or two hundredths. At this same setting the squareness of this hole with the base, of the casting can be checked in one direction with the height gauge by feeling along the underside of a protruding length of test bar. Its squareness in the other direction can be checked by verifying that a protruding length of test bar is parallel with the front machined face (B) of the holes (Fig. 38(c)).

(e) The remainder of the inspection consists of checking uniformity and dimensions of metal thicknesses, general overall dimensions and other sizes given on the drawing. All these may be finished to a sufficiently wide tolerance to allow of their being checked with the rule.

## Angular and taper testing

We have already discussed the use of the vernier protractor for measuring angles, and of taper gauges for checking standard tapers (Part I, pp. 207 and 342). For many jobs, however, we must employ additional methods to these.

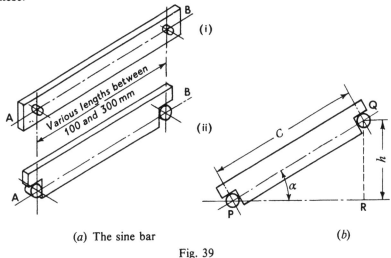

(a) The sine bar            (b)

Fig. 39

## The sine bar

For accurate work in connection with angles the sine bar possesses advantages over the protractor when conditions are favourable for its use. Sine bars differ in form, but the considerations affecting their setting are the same in every case. Two common types of sine bar are shown in Fig. 39(a). The bar shown at (i) has two plugs which are let in and project about 12 mm from the front face. At (ii) is shown a bar which is stepped at the ends, with a roller secured into each step by a screw which holds it in contact with both faces of the step. For the sine bar to be accurate the following points in its construction are important:

    (a) The rollers or plugs should both be of the same diameter.

(*b*) Their centre distance must be absolutely correct. (The bars available generally have centre lengths of 100 mm, 200 mm, 250 mm and 300 mm.)

(*c*) The centre line AB of the plugs must be absolutely parallel with the edge of the bar used for measuring (generally the bottom). It is desirable for the two edges of the bar to be parallel, and AB parallel with both.

When in use, the bar shown at (i) lends itself to clamping against an angle plate, whilst the one at (ii) can be rested on two piles of slip gauges to give it the correct inclination.

**Calculation for sine bar setting.** In Fig. 39(b), $C$ is the centre distance of the plugs, $h$ is the height of one plug above the other and $\alpha$ is the angular setting of the bar. Then

$$\frac{QR}{PQ} = \frac{h}{C} = \sin \alpha$$

*and* $$h = C \sin \alpha$$

i.e. difference in height of plugs = (centre distance)(sine of angle).

**Accuracy of the sine bar.** When using the sine bar on an angular surface it is important for the bar to be accurately set in the line of greatest slope, i.e. perpendicular to lines on the surface which are horizontal. Unless this condition is satisfied the bar will be inclined at some angle less than the one it is desired to measure (see Part I, Fig. 147).

Given an accurate sine bar and reliable means of setting its plugs to relative heights (e.g. slip gauges or a good height gauge), the degree of accuracy to which angles may be obtained depends on having a good set of tables of sines, and upon the magnitude of the angle. The larger the angle, the less is the possible accuracy of the setting. For example, from tables, the sines of various angles are as follows:

| Angle | 20° | 20° 1' | 40° | 40° 1' |
|---|---|---|---|---|
| Sine | 0·342 02 | 0·342 29 | 0·642 79 | 0·643 01 |
| Height difference for 200 mm bar (mm) | 68·40 | 68·46 | 128·56 | 128·60 |
| Angle | 60° | 60° 1' | 80° | 80° 1' |
| Sine | 0·866 03 | 0·866 17 | 0·984 81 | 0·984 86 |
| Height difference for 200 mm bar (mm) | 173·21 | 173·23 | 196·96 | 196·97 |

From this we see that the height difference variation for a 1' change of angle at 20° is $68·46 - 68·40 = 0·06$ mm and this becomes less as the angle increases until at 80°, a difference of 1′ in the angle makes only 0·01 mm difference in the relative height of the plugs. For this reason, as well as for avoiding high piles of slip gauges when these are used, the complements of angles above 45° should be tested if possible (see Fig. 42).

The degree of accuracy to which angles may be measured may be gathered from the table just given, in which it will be seen that for angles up to 40°, a difference of 1 minute in the angle causes a variation up to about 0·05 mm in the relative height of the ends of a 200 mm sine bar. This is sufficient to make possible the measurement of fractions of a minute if tables are available giving the sines of such divisions.

**Applications of the sine bar**

1. *To check the taper gauge shown at Fig. 40(a)*

Since the taper on the gauge is 1 in 10, the tangent of half its included angle will be $\frac{1}{2} \div 10 = \frac{1}{20} = 0\cdot050$ (see Fig. 40(b)), from which the included angle is 5° 44′, and assuming a 200 mm sine bar to be used, the height of one plug above the other will be

$$200 \sin 5° 44' = 200 \times 0\cdot099\,89 = 19\cdot998 \text{ mm}$$

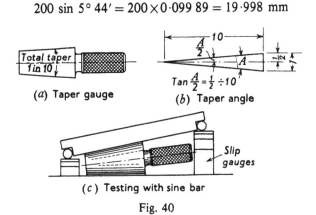

(a) Taper gauge          (b) Taper angle

(c) Testing with sine bar

Fig. 40

Two combinations of slip gauges must be chosen which will have the above differences when assembled, and which will raise the bar to a convenient height for accommodating the gauge. When the gauge is being tested care should be exercised to ensure that its centre is approximately under the centre of the bar and that it is in line with the length of the bar. The conformity or otherwise between the upper surface of the gauge and the sine bar may be checked either by the appearance of light when viewed against a bright source, or by interposing a cigarette paper at each end and trying each for tightness (Fig. 40(c)).

## 2. *To check the angle of the machined step shown at Fig. 41*

To do this the sine bar may be placed directly on the step and its heights measured from the surface plate with a height gauge or the casting may

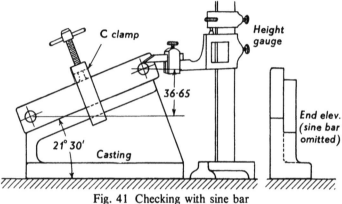

Fig. 41 Checking with sine bar

be tilted and packed up until the step is parallel with the surface plate, when the angle of tilt of the base may be checked with the sine bar. If the job is carried out according to the first plan the diagram of the set up is shown and, assuming a 100 mm sine bar, the height difference between its plugs should be

$$100 \sin 21° 30' = 100 \times 0.3665 = 36.65 \text{ mm}$$

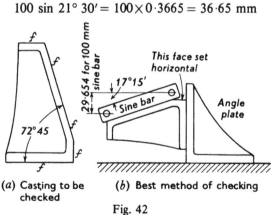

(a) Casting to be checked     (b) Best method of checking

Fig. 42

This may be checked with a height gauge as shown.

### 3. *To check the casting shown at Fig. 42(a)*

As we noted earlier, when the angle to be tested would involve a steep tilt of the sine bar the complement of the angle should be tested. To test this casting, bolt it to an angle plate of proved accuracy, and ensure that the narrow machined strip joining the base to the angular face is set horizontal. Place the sine bar on the inclined face as shown and set it square with the angle plate (Fig. 42(b)). The reading of the sine bar may now be taken, and since the original angle is 72° 45′, the angle to be tested will be $90° - 72° 45′ = 17° 15′$.

The sine of this angle is 0·296 54, so that the height difference for a 100 mm sine bar would be 29·654 mm.

### Position checks for angular surfaces

The most convenient method of making tests for the position of angular surfaces is by the employment of cylinders set in contact with the surfaces to be measured. The following examples will serve to show the principles of such methods, and suggest to the reader how other jobs of a similar nature might be tackled.

### 1. *To check the size of the groove shown at Fig. 43(a)*

In problems of this type it is impossible to measure with precision to angular corners, so that the only accurate method is to base the method of checking on the actual sloping sides of the vee. This is consistent with practical requirements since it is generally the vee sides which are important and not the corner extremities of them.

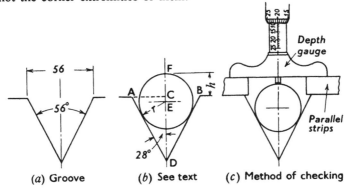

(a) Groove    (b) See text    (c) Method of checking

Fig. 43 Checking an angular groove

If a cylinder, in the form of a test bar or roller is placed in the vee, the conditions are shown at Fig. 43(b), and the mathematics of the problem are as follows:

If $r$ is the radius of the roller then

$$\frac{r}{ED} = \sin 28° \quad \text{and} \quad ED = \frac{r}{\sin 28°} = \frac{r}{0·46947} = 2·130r$$

In triangle ACD, $\quad AC = \frac{56}{2} = 28 \text{ mm}$

$$\frac{AC}{CD} = \tan 28° \quad \text{and} \quad CD = \frac{AC}{\tan 28°}$$

$$= \frac{28}{0·5317} = 52·66 \text{ mm}$$

Now $h = FC = DF - DC = ED + EF - DC$
$$ED = 2·130r; \ EF = r \quad \text{and} \quad DC = 52·66 \text{ mm}$$
$$\therefore h = 2·130r + r - 52·66$$
$$= 3·130r - 52·66 \text{ mm}$$

A suitable roller or test bar would be one of 38 mm diameter, then $r = 19$ and $h = 3·130 \times 19 - 52·66 = 6·81$ mm.

Before making the test the angle of the vee must be checked as well as its symmetry with the centre line, as if these are not correct the above analysis will not hold good. If these are in order, the plug may be placed in the vee and the above height from the top surface tested by a convenient method. One way, using parallels and a depth gauge, is shown at Fig. 43(c).

## 2. To check the dovetail slide form shown at Fig. 44(a)

As before, the 55° angles must first be tested for accuracy and then two test bars may be placed as shown at (b). A suitable test bar would be 19 mm diameter and referring to (c).

$\frac{AB}{CB} = \tan 27\frac{1}{2}°$ (since the centre of the bar will lie on the bisector of the angle), hence

$$CB = \frac{AB}{\tan 27\frac{1}{2}°} = \frac{9·5 \ (\text{rad of bar})}{0·520\,57} = 18·44 \text{ mm}$$

If the 95 mm dimension is correct, the reading to the centre of the two 19 mm test bars will therefore be $95 + 2(18·44) = 131·88$ mm. This may be checked by measuring inside the plugs with slip gauges, or over the outside of their diameters with micrometer or vernier (Fig. 44(b)).

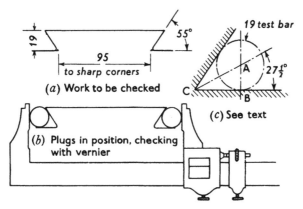

Fig. 44 Checking dovetail slide

### 3. To check the position of the sloping face in Fig. 45(a)

After verifying the accuracy of the 26° 30' angle, a 38 mm roller may be placed in the angle and the measurement made with slip gauges as shown at (b).

The calculation necessary to determine the distance to be measured will be seen from Fig. 45(c) and the following:

Since angle EAB = 26° 30', angle ABC will be

$$90° + 26° \ 30' = 116° \ 30'.$$

When the roller is placed in the angle its centre will lie on the bisector of angle ABC, i.e. angle DBC $= \dfrac{116° \ 30'}{2} = 58° \ 15'$

In triangle FBG, $\hat{G} = 90°$; $\hat{B} = 58° \ 15'$ and FG = 19 mm

$$\frac{FG}{BG} = \tan 58° \ 15', \quad \text{i.e.} \ \frac{19}{BG} = 1 \cdot 615 \ 98$$

from which BG $= \dfrac{19}{1 \cdot 615 \ 98} = 11 \cdot 76$ mm

Hence GC $= 52 \cdot 50 - 11 \cdot 76 = 40 \cdot 74$ mm

and the distance from the diameter of the plug to the face

$$= 40 \cdot 74 - 19 = 21 \cdot 74 \text{ mm}$$

### The checking of tapers

Because of the extreme difficulty of making accurate measurements to angular corners and edges, a reliable method of checking a taper for which a corresponding taper ring gauge is not available is to take measurements

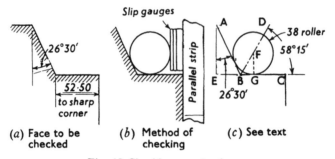

(a) Face to be checked

(b) Method of checking

(c) See text

Fig. 45 Checking angular face

over rollers or balls set in contact with its conical surface and from the readings obtained to deduce its accuracy. The angle of a taper may, as we have seen, be checked with a sine bar, but this method cannot give us the diameter at any particular point on its length.

The following examples will illustrate the principles of the method:

*To check the taper shown at Fig. 46 (a)*

In order to carry this out two sets of rollers are necessary which must be set in contact with the surface of the taper at two positions along its length, as far apart as possible. The centres of each pair of rollers must be on a line perpendicular with the axis of the taper, and diametrically opposite

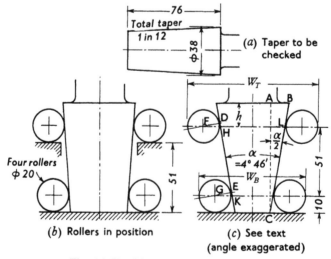

(a) Taper to be checked

(b) Rollers in position

(c) See text
(angle exaggerated)

Fig. 46 Checking a taper with rollers

rollers should be equal, but it is not necessary for upper and lower pairs to be of the same size. For convenience of working, the taper may be stood with one end on a surface plate and for this purpose the end to be used should be faced up flat, or slightly concave. If much work of this type has to be measured some form of fixture for supporting the taper and carrying the rollers is helpful, but for occasional jobs a set-up such as we have shown will serve. In this case we will stand the taper on its smaller end and use four 20 mm rollers, one pair resting on the plate and the others supported with their centres 61 mm above the base, as shown at Fig. 46(b). The distance over the outside diameters of each pair of rollers may be measured with micrometer or vernier and the relationships concerned are obtained as follows (Fig. 46(c)).

Since the taper is 1 in 12, the difference in the overall centre distance of rollers spaced at 51 mm along its length (i.e. $W_T - W_B$) will be $\frac{51}{12} = 4.25$ mm.

(Had the taper been given in the form of an included angle $\alpha$, this distance would have been $2\left(2 \tan \frac{\alpha}{2}\right)$. See diagram.)

We now have to find $W_T$ for a taper top diameter of 38 mm.

In triangle ABC, $AB = \frac{76}{24} = 3.17$ mm, $AC = 76$ mm, and if $\alpha$ is the included angle of the taper, then $\frac{AB}{AC} = \tan \frac{\alpha}{2}$

i.e.
$$\frac{3.17}{76} = 0.041\ 71$$

From this
$$\frac{\alpha}{2} = 2°\ 23'.$$

Referring to one pair of rollers, these will touch the taper at D and E, angles DFH and EGK each being equal to $\frac{\alpha}{2} = 2°\ 23'$, and in triangle FDH, $\hat{D} = 90°$.

Hence $\frac{FD}{FH} = \cos 2°\ 23'$ and $FH = \frac{FD}{\cos 2°\ 23'} = \frac{10}{0.999\ 14} = 10.011$ mm.

The distance $h$ from the line FHL to the top face of the gauge must now be measured with height gauge from the rollers, or by some other suitable method. Actually, if the length of the gauge is exactly 76 mm, this will be 15 mm, but as the length of the gauge is generally unimportant an exact determination is necessary.

When $h$ is known, the diameter HL can be calculated since

$$HL = 38 - \frac{h}{12}$$

We will assume $h$ to be 15 mm

Then $\quad HL = 38 - \dfrac{15}{12} = 38 - 1 \cdot 25 = 36 \cdot 75$ mm

and distance over top rollers

$= HL + 2FH + 2(\text{rad of roller}) = 36 \cdot 75 + 2 \times 10 \cdot 011 + 20 = 76 \cdot 77$ mm.

Distance over bottom rollers $= 76 \cdot 77 - 4 \cdot 25 = 72 \cdot 52$ mm.

(A full discussion of the checking of tapers in this way is given in the author's *Senior Workshop Calculations*.)

## Projection method of checking

A convenient and effective method of checking the profile and dimensions of small shapes is to magnify their profile by means of lenses and transmitted light, and to compare the enlarged image with an accurate layout made to the scale of the enlargement. Sizes may be checked by direct measurement on the enlarged shadow, and subsequent division by the multiplication factor. The essential features of a good projection apparatus are that the enlarged projection shadow shall be a faithful reproduction of the object, that the magnification factor is accurate and that the design of the apparatus permits of the maximum latitude in holding and adjusting the object, and examining the projected shadow.

Modern high-class projection equipment satisfies the above requirements to a high degree and represents a very efficient though simple means of examining work which would otherwise present great difficulties. The magnifications available usually vary from 10 to 50, and in some cases 100, a choice of, say, ×10, ×30, and ×50 being obtained on the same apparatus by fitting different lenses. Designs vary but essential details are a condenser or collimator for projecting a parallel beam of light from the light source on to the object which is located between the condenser and the projection lens. After passing through the lens the light may shine directly on to the receiving screen or be reflected on to it by one or more mirrors. Fig. 47 shows the principle of the Vickers Contour Projector in which light from the lamp-box passes through the condenser O, picks up the object to be projected and then passes through the lens J. It is then

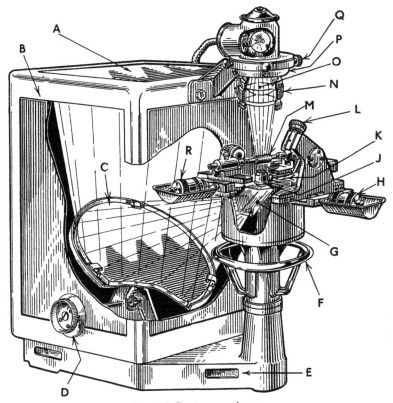

Fig. 47 Contour projector

deflected by the roof prism G on to the mirror C, which again deflects it on to the projection screen A. On the instrument, C and the light passing to it are enclosed by the casing shown broken. According to the lens being used this instrument enables magnifications of ×10, ×25 and ×50 to be obtained. Symbols R and H show micrometer heads for moving the stage upon which the projected object is carried so that minute movements may be made for the purpose of comparison on the screen. The symbol L is a micrometer adjustment for swinging the stage to an angle, an adjustment necessary when screw threads are being examined as the light must pass through the thread along its helix angle.

Projection methods of examination are well adapted to the examination of form tools, profile gauges, press-tools, gear teeth, screw threads, etc., and the projected shadow, compared with a master profile or measured, provides an efficient method of controlling the shape of intricate forms. For examination of the common screw thread forms templates may be obtained, accurately made to shape and magnification. For checking other classes of work, in addition to micrometer operated stages of the type

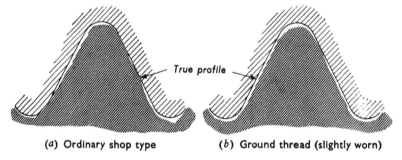

(a) Ordinary shop type        (b) Ground thread (slightly worn)

*Wolverhampton Polytechnic*

Fig. 48 Projected profiles of tap threads

shown, slip gauge holders, protractors and other accessories may be used. The projector shown has a screw thread in position, but flat and other shaped articles are just as easily accommodated.

An impression of the appearance of projected images and their master layouts is given at Fig. 48, which also provides an interesting study as to the value of a ground thread tap when accurate threads are important.

# 3   Machine tools

A machine tool is a power-driven apparatus designed to perform certain metal-removing operations and produce a desired form on the surface being machined. As most machined surfaces require to be either cylindrical or flat, the work of machine tools in general is concerned with producing one or both of these forms. The design of the machine incorporates means for holding and supporting the work and the cutting tool, in addition to which arrangements are made for the necessary movements of work and tool to produce the desired effect. For example, when turning, a lathe holds and rotates a piece of metal and at the same time causes a cutting tool to move on a line parallel with the axis of rotation. The result, when the tool has completed its cut, is the formation of a cylinder and the nearness of this cylinder to perfection is dependent on the accuracy with which the lathe performs its movements and upon such factors as its rigidity, condition, etc.

Machine tools may be of the *standard*, or *special purpose* variety. A standard machine is one which is able to deal with a variety of work and permits of a reasonably wide range of operations to be performed. A special purpose machine, on the other hand, is one which has been designed for some specific purpose and only performs one, or a limited range of operations; its application often being confined to a single variety of component. A centre lathe is a good example of a standard machine tool, for upon it we are able to perform a wide range of work, including the following: 1. turning between centres; 2. turning, facing, drilling and boring in the chuck; 3. turning, boring and facing work held on the faceplate; 4. boring work held on the carriage using a boring bar held in the spindle. If we wanted a machine purely for such an operation as facing the heads of bolts we could design one something like a lathe, but it would be much simpler, and would not require to have a carriage, compound slide, or tailstock. The headstock would need to be fitted with some form of permanent chuck and probably only a single speed would be necessary for the spindle. A special purpose machine such as this would face bolt heads much more efficiently than they could be done on a standard lathe, but it would lack the versatility of the lathe for a wide range of work. Single purpose machines are not always as simple as our above discussion might suggest, but in most cases they perform operations which could be done on one of the standard machines, but for producing large quantities the single purpose machine has the advantage in efficiency on that one job, because the standard machine, designed for reasonable efficiency over a range of work, cannot show high efficiency when applied to one particular operation.

As the scope of our present work is a treatment of the fundamentals of workshop technology we shall confine ourselves to the chief types of standard machines, but the reader, if he understands the principles of these, should have no difficulty with most of the special purpose machines since they are often developments of standard machines. The chief standard machines in addition to the lathe are: 1. milling, 2. drilling, 3. shaping, 4. planing, 5. boring, and 6. grinding machines. We propose to discuss each of these in subsequent chapters, reserving the remainder of this chapter for the consideration of machine tools in general, since all machines, whatever their type, have certain common features.

## Elements of construction

### The bed, or body, or frame

The main framework of a machine, generally an iron casting, forms the fundamental of its construction and carries all the other components included in the design. In the case of the lathe, planer, boring machine and grinding machine, this casting is called the *bed*, whilst the main casting of the shaper and miller is called the *body*. The lower surface of this casting is un-machined and rests on the floor, being provided with lugs and holes for bolting the machine down. The bed must be of solid construction for a base, and so that it may serve as a buffer for absorbing the vibrations set up when the machine is at work. It must be rigid because the forces brought into play by the cutting operation are transmitted to the bed and if these caused it to distort unduly the machine would lose its accuracy. It is possible for a bed to distort under its own weight if carelessly installed on an unlevel floor. Certain parts of the bed are machined so that accurate location may be provided for other components included in the construction, and a feature which is common to most beds are the 'ways', which are long machined strips or grooves for carrying and guiding a slide or table. In

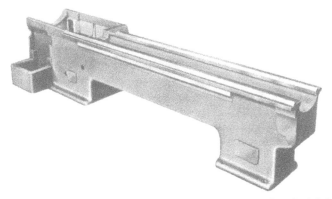

*Dean Smith & Grace Ltd*

Fig. 49 Bed of lathe

addition to the main ways or guides, the machine bed will have machined facings for the purpose of attaching various components, e.g. feed motion, driving motor, pump, etc., and, in the case of some machines, the bosses and bores for carrying the main spindle or some part of the drive may be incorporated in the bed casting. Such occurs on the milling machine where the main spindle is carried directly in the body casting, and on some planers where the drive enters at the side of the bed.

The bed of a lathe is shown at Fig. 49, and the body of a milling machine at Fig. 50. We might, with the help of these two diagrams, bring to the reader's attention one or two important fundamentals which are of vital concern to the accuracy of the machines. On the lathe bed the front Λ and the rear flat ways support and guide the saddle whilst the rear Λ and the inner front flat ways carry the tailstock. The headstock is carried on the facing at the left end of the bed. For this machine to produce accurate work the following conditions are essential:

(*a*) Axes of Λ ways straight and parallel in vertical and horizontal planes and parallel with the flat ways serving as part of the support. These flat ways should be truly linear.

(*b*) Axes of Λ and flat ways parallel with the horizontal and vertical facings which serve to locate and support the headstock.

The reader, when he has read our section on alignments, should decide for himself why these conditions are so necessary and their influence on the accuracy of the work produced. There are diagonal cross ribs in the

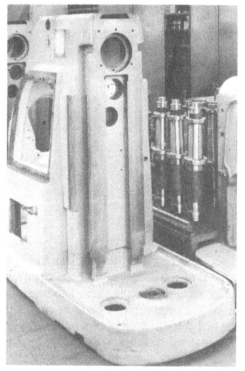

*Cincinnati Milling Machines, Ltd*

Fig. 50 Body of a milling machine

centre of the bed for stiffening purposes which serve to minimise distortion and twist of the bed under the cutting forces.

In the milling machine body (Fig. 50) the vertical ways at the front guide and locate the knee of the machine with which is incorporated the cross-slide, longitudinal table and slide, and table swivelling arrangement (if machine is universal). Everything, therefore, concerning the location and movement of the work is primarily dependent on the accuracy of these vertical ways. In milling it is important that the cutter axis shall be parallel and perpendicular to the two work location axes. This alignment is dependent on the spindle which in its turn depends on the two bores for receiving it. An important requirement of the milling machine body, there-fore, is absolute squareness between the vertical knee-way faces and the spindle bore axis.

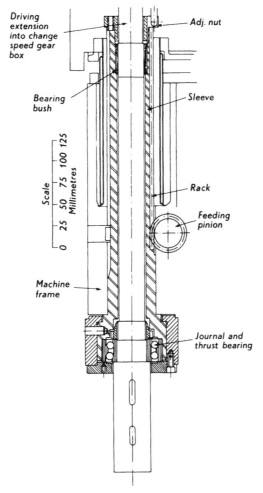

Driving extension into change speed gear box

Adj. nut

Bearing bush

Sleeve

Scale Millimetres

25  50  75  100  125

0

Rack

Feeding pinion

Machine frame

Journal and thrust bearing

*Fredk. Pollard & Co, Ltd*

Fig. 51(a) Spindle for 530 mm upright drill.

## The main spindle

On the machines we are considering, with the exception of the shaper and planer, the spindle serves an important purpose in the functioning of the machine. The spindles of the lathe and vertical boring mill support and drive the work, whilst in the case of milling, drilling, horizontal boring and grinding machines the spindle carries the tool (e.g. drill, cutter, grinding wheel, etc.) which operates on the part being machined. There is no main spindle function on the shaper and planer because, unlike the other machines which employ a rotatory motion for the removal of material, these machines do so by an oscillatory motion of the tool (shaper) or work (planer). There are, of course, driving shafts or spindles on shapers and planers, but these do not exercise any direct influence on the accuracy of the work turned out. The spindle is supported in bearings which are in some way located and clamped in bores or housings incorporated in the machine body (miller), headstock (lathe), or, for the other machines, in a spindle head of some form suitable to the general design of the machine. The bearings must locate the spindle so that its axis of rotation coincides with the axis of the bores in which the bearings are located, and they must support and maintain the spindle on this axis against the disturbing pressure of the forces caused during cutting. In addition to side forces, a spindle may be subjected to thrust forces along its length (e.g. when drilling work held in the lathe chuck), so that as well as requiring supporting (*journal*) bearings a spindle may need to have *thrust* bearings to absorb end forces. The bearings used for the journal support vary, some makers preferring plain bronze or white metal, whilst others rely on ball and/or roller bearings. For the support of light thrust loads a collar on the shaft bearing

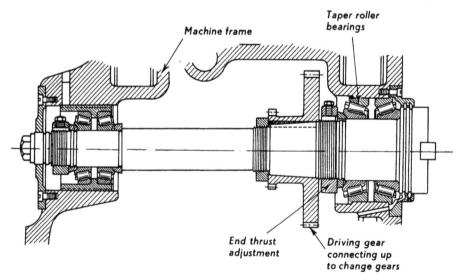

Cincinnati Milling Machines, Ltd

Fig. 51(b) Milling machine spindle

against a suitable bush head on the frame may be sufficient, but for heavy and continuous work ball thrust races are used. The working end of the spindle is invariably bored with a taper hole into which are fitted the various accessories employed. Diagrams of lathe spindles with plain and roller bearings are shown at Fig. 69(b), and cross-sections of drill and milling machine spindles are given at Figs. 51(a) and (b).

**Spindle driving**

The drive to a spindle may be by belt direct to a pulley on the spindle, by gearing from a pulley or chain wheel attached to another shaft or by some form of gear connection from a pulley situated at a point convenient for the design of the machine. When the spindle is only required to run at one set speed (e.g. a grinding machine) a single pulley attached to the shaft is sufficient, but when a range of speeds is necessary, means must be pro-

*Cincinnati Milling Machines, Ltd*

Fig. 52 Enclosed motor drive to milling machine

vided to obtain them. For lathe and milling machine spindles the most common form of drive is an all-gear arrangement. An example of this is shown in Fig. 94 in W.T. Part I. The single input pulley of an all-gear head may be driven by belt from an overhead shaft or by vee belts from a motor attached to the machine. Before the development of the all-gear drive, machines were driven by a flat belt through a cone pulley and back gear, but the cone drive is now restricted to light, simple drives, an application to a bench drill being shown in W.T. Part I, p. 137, Fig. 93. The enclosed drive to a self-contained milling machine is shown at Fig. 52. In the case of drilling and horizontal boring machines it is often more convenient to locate the motor or pulley and gearbox at some distance from the spindle and to employ gears or driving shafts, or some other form of transmission, for conveying the power. An example of the drive from a remote point is shown diagrammatically at Fig. 53, which illustrates the method used on the Kearns 'S' type horizontal boring machine. Power from the motor is

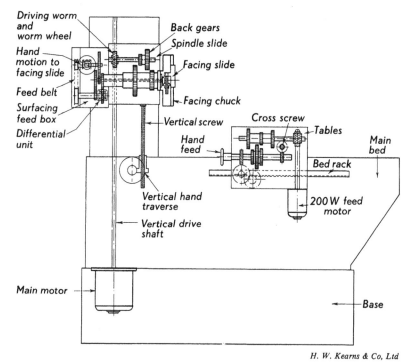

*H. W. Kearns & Co, Ltd*

Fig. 53 Drive to the spindle of the Kearns 'S' type horizontal boring machine

taken by a vertical drive shaft and through a worm and wormwheel to the change gearbox and thence to the spindle. The diagram also shows the method of actuating the facing slide and of feeding the table from a 200 W feed motor.

**Spindle speeds**

The machines with which we are dealing, being of a general purpose character, must be able to undertake all sizes of work within their range. For example, a lathe which will accommodate work up to 400 mm diameter must have a range of speeds suitable for a range of diameters from about 20 mm to 400 mm. In the same way a drilling machine designed to drill holes from 12 mm to 32 mm must have suitable speeds for all drill sizes within that range.

Generally, when fixing the speed range for a machine, the smallest and largest cutter or work sizes and the cutting speed to be used determine the highest and lowest speeds.

*Example* 1. *Determine suitable highest and lowest spindle speeds for the lathe above, assuming a cutting speed of* 20 *metres/min.*

and the lowest speed $S = 20$, $d = \dfrac{400}{1000}$

The formula connecting rev/min and cutting speed is $N = \dfrac{S}{\pi d}$

where $N =$ rev/min of spindle
$S =$ Cutting speed metre/min
$d =$ Work or cutter diameter (metre)

For the highest speed $S = 20$ and $d = \dfrac{20}{1000}$

Hence $\qquad N = \dfrac{20}{\dfrac{22}{7} \times \dfrac{20}{1000}} = \dfrac{7000}{22} = 320$ rev/min

At the lowest speed, $S = 20$ and $d = \dfrac{400}{1000}$

Hence $\qquad N = \dfrac{20}{\dfrac{22}{7} \times \dfrac{400}{1000}} = \dfrac{7000}{22 \times 20} = 16$ rev/min

When the limiting speeds have been decided upon, the total number of speeds in the range must be known and the intermediate speeds can be fixed. The usual method of doing this is to arrange it so that the series of speeds form a geometric progression and the method of working this out is as follows:

*Example 2. If the above lathe is to have 8 speeds in geometric progression, determine the speeds.*

A geometric progression is a series of quantities in which each succeeding term is obtained by multiplying the previous term by a certain fixed amount.

Thus:

| | 1 | 2 | 4 | 8 | 16 | 32 | etc. |
|---|---|---|---|---|---|---|---|
| | 1 | $\frac{1}{3}$ | $\frac{1}{9}$ | $\frac{1}{27}$ | $\frac{1}{81}$ | $\frac{1}{243}$ | etc. |
| | 16 | 24 | 36 | 54 | 81 | $121\frac{1}{2}$ | etc. |
| and | $a$ | $ar$ | $ar^2$ | $ar^3$ | $ar^4$ | $ar^5$ | etc. |

are all examples of geometric progressions.

In our case we have to find a geometric progression having eight terms of which we know the first and last, and this we can do if we can find the multiplier for operating on each term to get the one succeeding. Let us call this multiplier $r$, and write the series down ([1st term = 16 and 8th term = 320).

| Term No.. . . . | 1 | 2 | 3 | 4 | 5 | 6 | 7 | 8 |
|---|---|---|---|---|---|---|---|---|
| Value . . . . . | 16 | $16r$ | $16r^2$ | $16r^3$ | $16r^4$ | $16r^5$ | $16r^6$ | $16r^7$ |
| | | | | | | | | (320) |

(See the fourth series above.)

But since the 8th term must be 320:

$$16r^7 = 320 \quad \text{and} \quad r^7 = \frac{320}{16} = 20$$

From which $r = \sqrt[7]{20}$ giving $r = 1.534$.

We may now calculate the whole range of speeds:

| | | | | | |
|---|---|---|---|---|---|
| 1st speed = 16 rev/min | | | 5th speed = $16 \times 1.534^4$ = 89 rev/min | | |
| 2nd ,, | = $16 \times 1.534$ = 25 rev/min | | 6th ,, | = $16 \times 1.534^5$ = 136 ,, | |
| 3rd ,, | = $16 \times 1.534^2$ = 38 ,, | | 7th ,, | = $16 \times 1.534^6$ = 210 ,, | |
| 4th ,, | = $16 \times 1.534^3$ = 58 ,, | | 8th ,, | = $16 \times 1.534^7$ = 320 ,, | |

and the full range is

| Speed No.. . . . | 1 | 2 | 3 | 4 | 5 | 6 | 7 | 8 |
|---|---|---|---|---|---|---|---|---|
| Rev/min . . . . | 16 | 25 | 38 | 58 | 89 | 136 | 210 | 320 |

(As an additional example, the reader might determine the most suitable diameter to be turned on each speed and make a table showing Speed No.—Work dia, which could be used by the machine operator.)

It might be wondered why the speeds should be graded in a geometric progression and not in equal steps (arithmetic progression). To answer this we have drawn a graph on which the speeds are represented, and it will be seen that the geometric relation gives a curve, whilst the arithmetic range appears as a straight line (Fig. 54). In each case the 1st and 8th speeds are

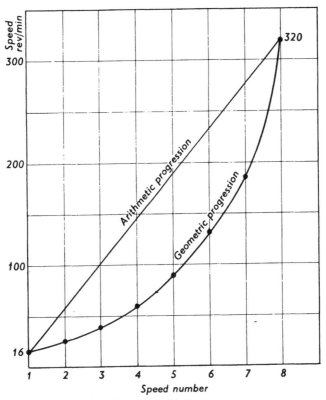

Fig. 54 Graph of lathe speeds

16 and 320, but a comparison of the 2nd shows it to be 25 on the geometric against 58 on the equal step method. Now the jump from 16 to 58 rev/min would be far too great at the bottom of the range and would mean that no speed was available for any work between approx. 400 mm diameter (16

rev/min) and 110 mm diameter (58 rev/min). In an opposite sense at the high end of the range the 7th speed on the equal step method would be 275 rev/min, which is too close to 320 rev/min to be of any useful alternative value. The geometric method, as the graph shows, reduces the speeds sharply at the high end and slowly at the bottom, giving a range which is much more suitable for practical requirements.

### Arrangements for carrying work

As we have already discussed the means of work holding used on the lathe*, we may proceed at once to the other machines under consideration. With the exception of grinding, all the other machines are to some extent uniform in their arrangements insomuch that a flat table forms the base upon which the work, or its holding fixture is bolted. Generally, the table surface is provided with tee slots for accommodating the heads of holding-down bolts and the slot sides provide the means for lining up a surface parallel or perpendicular to important axes (e.g. spindle) on the machine. When, for the purpose of cutting, the work must be fed along, the table is provided with a feeding lead-screw and, as we shall discuss later, this may be con-

*James Archdale & Co, Ltd*

Fig. 55 Table, cross-slide, knee unit of milling machine

A, Vertical slides; B, Knee (Cross-slides on upper surface); C, Table on longitudinal slides; D, For cross movement; E, For elevating.

* *Workshop Technology*, Part 1

nected to a power-driven feed mechanism. Often it is necessary to be able to feed work in two directions at right-angles and to allow for this the table, with its slide, is mounted on a second set of slides running at right-angles to the table. On the milling machine the table moves longitudinally on its own slides under the operation of the table lead-screw. This is mounted on the cross-slide, the screw of which is operated from the front of the machine and the whole is given an up-and-down motion by being carried on the knee, moving on the vertical slides to which we have already referred. Fig. 55 shows the table, cross-slide and knee assembly of a vertical milling machine on which the slides, lead-screw handles and elevating handle are indicated. The elevating screw for the knee is underneath and is connected to the handle by a pair of bevel gears. The table has three tee slots running throughout its length and their faces are machined parallel with the longi-tudinal slide, so that they are perpendicular with the spindle axis (on a horizontal machine) and parallel with the front surface of the vertical knee slides. Vises, fixtures, etc., when bolted to the table, are lined up from the centre tee slot by means of tenon pieces which fit in the slot. To set faces of other work they may be pressed into contact with blocks fitting into a slot or set parallel, by scribing block or height gauge, direct from the front face of the vertical slides.

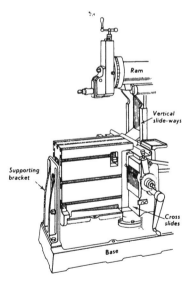

Fig. 56 Shaping machine table

The table of the planing machine has guides on its underside which slide on the ways of the bed. This table has longitudinal motion only, since the cross and vertical movements are provided on the rail which carries the tool. The table tee slots are parallel with its guides, so that the slots may be used for setting faces when parallelism with the travel is necessary. As a contrast to this, the shaping machine table has cross-wise and vertical motion only, the longitudinal movement necessary being made by the tool itself, via the machine ram. The vertical table movement is used mainly for adjusting the height to accommodate different jobs, and seldom for traversing a cut. This function is performed by the vertical slide attached to the end of the ram, and carrying the tool box. For its cross motion the table is carried on slides running across the front of the machine body, and is moved by a lead-screw which terminates at a turning handle on the operating side of the machine.

The table itself is of box form, being provided with the slots on top and sides, the sides being useful for holding certain classes of work. On the more elaborate machines the table may be swivelled to any angle by rotation about an axis parallel with the movement of the tool, a feature which adds greatly to the usefulness of the machine on angular work. A leg or bracket is usually provided at the front of the shaper table to prevent undue deflection of the unsupported end.

A shaper table is shown at Fig. 56.

**The machine vise**

An important auxiliary work-holding method used on shaping, milling and planing machines, and to a lesser extent on surface grinders is the machine vise. The general construction of this will be seen from Fig. 57, which shows a milling machine type at (a) and at (b) a model suitable for use on a shaper. The base bolts to the machine table and is located by a pair of tenon pieces which fit in the tee slot. The upper portion is joined to the base by a turntable graduated in degrees, and the tenon pieces locate the base so that at the extremities of the angular scale the jaws are parallel and perpendicular with the principal axes of the machine. For angular work the upper portion of the vise may be swivelled and locked. The sliding jaw is moved by a square or acme thread, and to promote accuracy and minimise lifting it is provided with a long bearing surface. Generally the body and base of the vise are of cast-iron and to allow for renewals the jaws are of steel, held in position by screws. On the smaller vises the jaws are hardened, but the large shaping vises have soft steel jaws. As the lowness at which

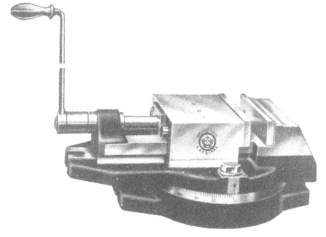

*The Brook Machine Tool Co*

Fig. 57(a) Milling vise

*The Butler Machine Tool Co, Ltd*

Fig. 57(b) Shaping machine vise

work is supported leads to increased rigidity vises are made as squat as possible. To promote this feature still further some milling vises are made so that the upper portion can be separated from the base and bolted direct to the machine table (see Fig. 57(a)). When work has to be machined which necessitates obtaining angles in more than one plane a *universal vise* can be used (Fig. 57(c)), which lends itself to a wide range of positions since a swivelling motion is provided both on the base and on the upper surface of the adjustable angle plate portion.

**Tightening the vise.** It is not generally necessary to use a hammer on the handle to tighten the vise sufficiently for its work. A good, steady and increasing pressure applied with both hands and the weight of the body will usually do all that is required. Blows with a steel hammer mutilate the

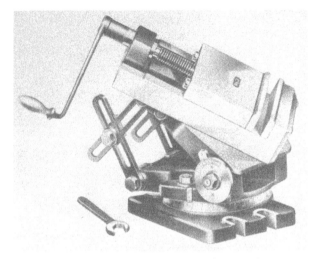

Fig. 57(c) Universal vise

handle disgracefully, whilst the harmful effects of a lead or copper hammer cannot be estimated and may only be discovered too late when the vise is fractured. An intensity of grip on the job greater than that required to support it against machining might result in distortion, so that when it is taken out the spring back to its original shape will destroy the machining accuracy. If the vise handle is broken and too short to be tightened by hand, attention should be called to the fact so that a new one may be fitted.

**The clamping of work**

In operating machine tools it is frequently necessary to clamp work, in vises, to tables, faceplates, etc., and this operation, simple as it may appear, is often worthy of consideration if it is not to spoil the result of an operation or even the job itself. In the first place, if the outer end of the clamp is supported on packing this should be at the same level as the surface being clamped preferably at a little greater distance from the bolt (Fig. 58(a)). The packing should be steady on a firm, flat base. The clamp itself should be thick and strong enough not to bend when the nut is tightened, and a washer should always be used under the nut. If these precautions are observed a reasonable pressure on a standard length of spanner will be all that is required; it should never be necessary to strain oneself or use tubes or hammers on spanners for the purpose of clamping.

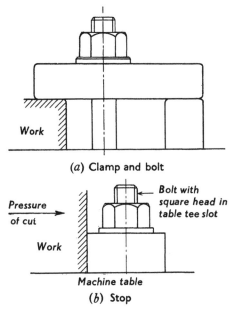

(a) Clamp and bolt

(b) Stop

Fig. 58 Clamping

It is easily possible, with too much tension on clamping bolts, to warp a machine table by the spring introduced. Whenever possible, the downward pressure of clamps should be helped in its function of preventing a job sliding by putting a stop as shown at Fig. 58(b).

The position at which clamping is applied is important, and if the surface is sloping in such a way that the clamp might slip it should be avoided if possible. When clamping on machined surfaces it is advisable to interpose a thin piece of copper or brass between the clamp and the surface if marking by the clamp will be detrimental to the product. An important final point is that concerned with the position of clamping and the support under the pressure. If there is solid metal between the clamp and the table the clamp may be tightened with confidence, but if this is not so caution should be observed to avoid distortion. If in doubt the clamps should be moved to a more suitable spot or packing introduced. Examples of this are shown at Fig. 59.

**Attachment of tools**

Excluding the grinding machine, which we shall discuss later, the means adopted for attaching and supporting the tool or cutter may be broadly

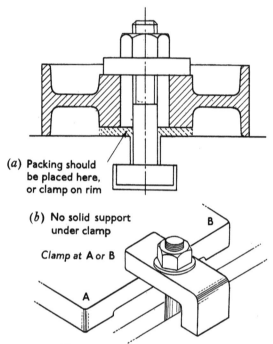

(*a*) Packing should be placed here, or clamp on rim

(*b*) No solid support under clamp

Clamp at A or B

B

A

Fig. 59 Unsupported clamping

divided into two main classes: 1. where a solid, single point tool is used (lathe, shaper and planer); 2. where rotary cutters and bars are used (milling, drilling and boring).

In the first case the most common method of holding and supporting the tool is by means of a tool box or tool-post, and examples of these are shown at Fig. 60. At (a) is the American type tool-post as applied to a lathe, in which the tool is clamped by the single centre screw and supported with its base on a block which has a rounded seating. By rocking this block on its seat, adjustment in tool height is obtained. The toolholder shown at (b) gives a more solid base for the tool, but packing must be put under the tool to vary its height. Tool changing and setting time is reduced by the use of a square tool-post of the type shown at (c) which holds four tools, and can be rotated into any of the four positions.

A shaping machine tool box is shown at Fig. 60(d). The principal feature of planer and shaper toolholders is the hinging of the tool carrying portion which permits the tool to lift and ride over the work on the non-cutting return stroke. On the diagram the hinge pin is indicated at A,

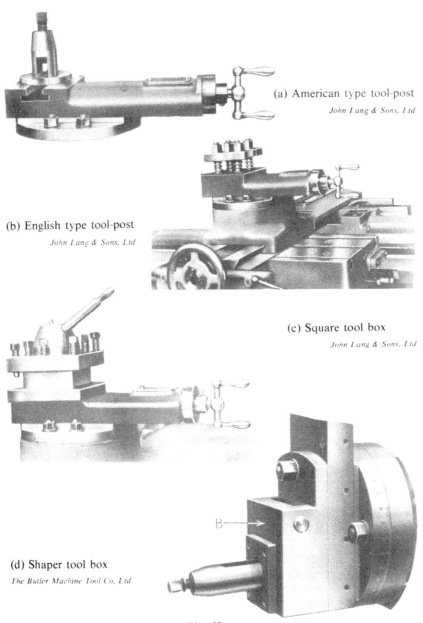

(a) American type tool-post

*John Lang & Sons, Ltd*

(b) English type tool-post

*John Lang & Sons, Ltd*

(c) Square tool box

*John Lang & Sons, Ltd*

(d) Shaper tool box

*The Butler Machine Tool Co, Ltd*

**Fig. 60.**

and when the table returns, the tool point, in fouling the surface of the work, is permitted to lift by the portion of the box marked B swinging out a small amount.

For securing and locating rotary cutters, drills and bars to the machine spindle the most common method is by means of a taper shank fitting into a taper hole in the spindle. The taper may be the end of an arbor or boring bar to which the cutters are attached and which, for purposes of rigidity, may be supported at other points in addition to its taper end. The taper, however, constitutes the main source of support and transmits the drive. A taper shank should be a perfect fit in its socket and if this condition is satisfied the friction of the fit is sufficient to transmit the drive in most cases.* Generally, however, an additional positive driving arrangement is provided, such as the tang of a drill or the driving dogs shown on Fig. 62, which fit into the two slots machined in the arbor flange. The taper shanks

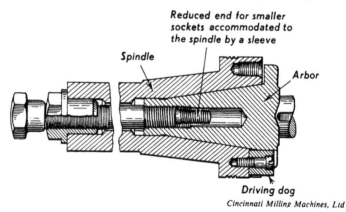

Reduced end for smaller
sockets accommodated to
the spindle by a sleeve

Spindle

Arbor

Driving dog

*Cincinnati Milling Machines, Ltd*

Fig. 61 Drawbolt (milling machine)

on heavier drives are secured by a drawbolt as shown at Fig. 61, which prevents any possibility of the taper slipping out and serves to introduce some initial tension on the fit of the taper. The spindle end of a milling machine is given at Fig. 62 which shows No. 50 of the group of tapers to the standard* agreed upon between British and American machine tool makers in 1937. The diagram of the cutter arbor in position at Fig. 63 shows, also, the additional support provided for the outer end of the arbor. For large facing

---

* The standard milling machine taper (Fig. 62) is an exception, the angle being too blunt for the taper to be self-locking and necessitating driving dogs (BS 1660, Part 3, 1953).

† BS 1660, Part 3, 1953.

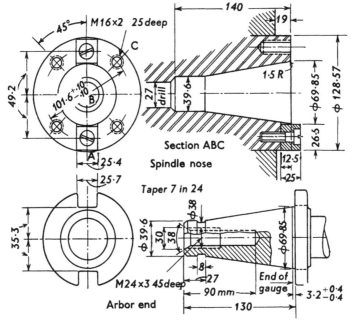

Fig. 62 Spindle nose of milling machine (Dimensions in mm)

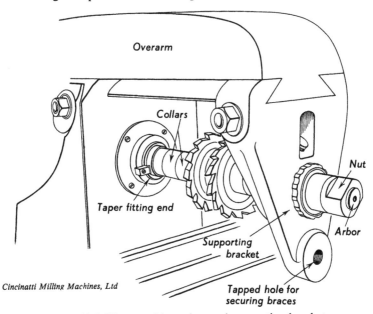

*Cincinatti Milling Machines, Ltd*

Fig. 63 Milling machine arbor and supporting bracket

cutters the spindle end shown is flanged and provided with tapped holes so that the cutter body may be spigoted and bolted direct to the flange. This avoids the use of a separate taper shank and gives increased rigidity.

On the horizontal boring machine the cutters are fitted to a boring bar which fits, with its taper end, into the machine spindle, and is supported by a bracket at its free end. A selection of boring bars is shown at Fig. 64.

*George Richards & Co, Ltd*

Fig. 64 Boring Bars

The tapers usually employed for the purposes we have just discussed are the Morse, or the Brown and Sharpe. Particulars of Morse tapers are given in Part I, Appendix 7, whilst Brown and Sharpe tapers are given in Appendix 2 of this volume.

### Speed and feed

Before we discuss feeding arrangements it will be well to be clear as to the speed and feed aspect in machine-tool operation. In order to machine and cover a surface there must be relative cutting motion between tool and work (speed), and then an additional movement must take place to enable the moving tool to cover the whole surface (feed). Thus, if a lathe or shaper is started up and the tool fed in until it cuts, it will merely make a groove, and no finished surface will be produced until an additional motion takes place to enable the tool to pass over the surface.

**Feeding arrangements.** In view of the above remarks, an essential feature of every machine tool is provision for feeding the tool or cutter over the surface it is machining, and on all but the very simplest of machines the principal movements are fed by power taken off the spindle or incoming drive by means of belt, chain, friction or gear drive. Many of the more modern and advanced machines have separate arrangements for the feeding which are electrically or hydraulically operated, but it is unlikely that we shall have the space, in this volume, to deal with these methods. Power-feed is not always applied to every movement on a machine, partly because of the cost and constructional difficulty of application and partly because certain of the slides are not often used for sustained movement during cutting, being provided more for the purpose of adjustment. Two examples of such movements are given by the compound slide of a lathe and the table of an upright drill. The compound slide is only used for feeding a cut when it is swivelled to some angle for the purpose of turning a chamfer or short taper, and for such purposes hand-feed is deemed to be sufficient, especially in view of the difficulty and extra cost that would be incurred in arranging for its power-feed. In the case of the drilling machine table it is only necessary to move it for varying the height to accommodate different work under the drill; the drill spindle moves to accomplish the cutting feed.

The relative motion necessary between tool and work for the purpose of traversing the surface being machined may be accomplished by feeding the tool over the work or vice versa. For example, on the lathe the work rotates at the cutting speed and the tool moves over it for the purpose of the traverse, whilst on the shaping machine the tool moves at the cutting speed and the work performs the feeding movement.

The methods used for obtaining the feeds on the machines we are considering vary widely, and we shall only be able to discuss representative types. When he has an opportunity the reader should study for himself, on the spot, the means used for feeding machine-tool elements, as many interesting and instructive mechanisms will be discovered. A combination which is common to many feeding arrangements is the screw and nut, and the function of the feed drive becomes that of slowly rotating the screw or nut in the required direction. As a means of imparting motion the screw and nut is convenient because, when the feeding mechanism is disengaged, the screw may be turned by hand for purposes of adjustment and for hand-feeding on short travels. The combination, however, suffers from one disadvantage: that of its positive nature. Unless some slipping or safety arrangement is incorporated in the drive serious damage may be caused to the mechanism if the slide, with the feed-motion engaged, is accidentally allowed to feed up to some solid part of the machine. Except for this disadvantage the screw and nut forms a satisfactory and convenient method in cases where its use is convenient.

**The milling machine.** A diagrammatic sectional sketch showing the feed-motion for moving the table and cross-slide of a horizontal milling machine

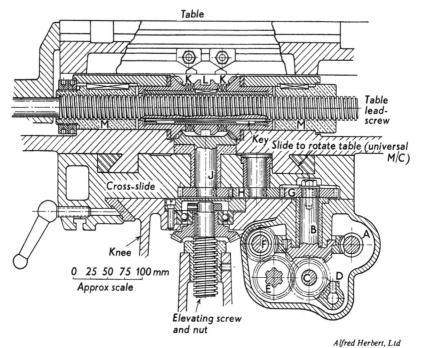

Fig. 65 Diagram of mechanism for feeding table and cross-slide of milling machine (see text). (Misc. details omitted or modified from original.)

is shown at Fig. 65. Shaft A is the power intake shaft from the column gearbox and a worm cut on it drives the wormwheel with which it is seen to be in mesh. This rotates shaft B which delivers the drive to the cross-slide by the bevel gear cut on its enlarged lower end, and conveys the table feed drive through gear G keyed to its upper end. For the cross-feed the bevel gear on B drives shaft C through another bevel. Another gear on C, slid longitudinally by fork D, can be moved into mesh with a sliding gear on shaft E which is fixed to the cross-slide proper and geared by a further pair of wheels to the cross-slide lead-screw F. The table feed drive is conveyed through G and H to shaft J, and from the bevel gear on its upper end to the bevels K which rotate in opposite directions. The clutch member L is a sliding keyed fit on the long sleeve which is also a sliding keyed fit on the table lead-screw. The feed engagement lever moves L to one side or the other, engaging the dog clutches by which L and the lead-screw are caused to rotate. This effects movement of the table as the screw engages with the

two nuts M. It will be seen that the left hand nut M can be moved longitudinally by lock-nuts. This feature is provided for the purpose of
adjustment to take out backlash from the motion.

**The vertical drill.** The feed mechanism of a vertical drill is shown at Fig.
66, and in this case the motion is imparted to the spindle by a pinion,
engaging in a rack cut on the spindle sleeve. Motion is taken by shaft A
from the main gearbox and cluster gear B, provided with a sliding feather
key, is actuated along the shaft by the lever C, which is operated by a
change lever on the front of the machine. The 3 gears on B provide three
changes of speed when each of them meshes with the corresponding gear
on shaft D. The bevel pinion E at the lower end of D drives the wheel F

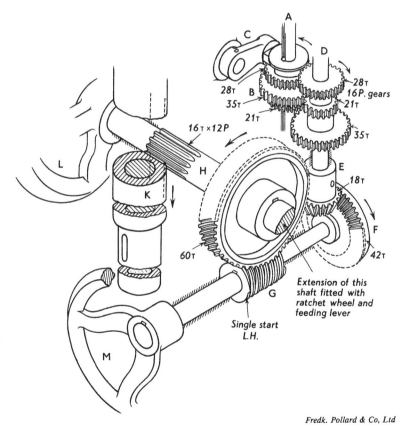

*Fredk. Pollard & Co, Ltd*

Fig. 66 Feed mechanism for upright drilling machine (Machine shown at Fig. 206)

and the worm G drives the wormwheel on shaft H which, with its gear teeth, operates the spindle sleeve K through rack teeth cut on the sleeve. L is a handwheel for rapid movement of the spindle when G is disengaged from the worm, and M is a handwheel for slow feed when B is disengaged. The right-hand end of shaft H (shown broken) is extended on the actual machine and is fitted with a ratchet wheel and pawl, with handle for hand-feeding. To disengage the feed the shaft carrying F, G and M is mounted on a pivot and may be swung down to throw G out of mesh with the worm-wheel. The machine to which this mechanism is adapted is shown at Fig. 206 so that the reader may pick out the corresponding points on that diagram.

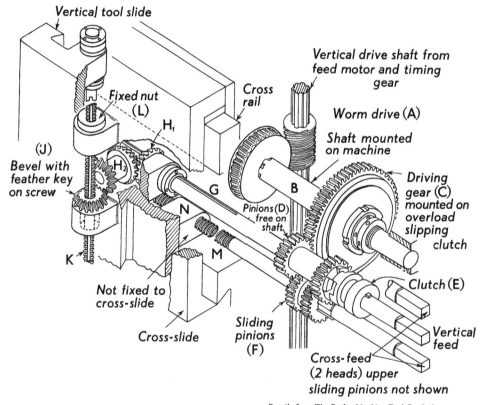

*Details from The Butler Machine Tool Co, Ltd*

Fig. 67 Diagrammatic illustration of the feeding mechanism of a planing machine. (Certain details omitted or modified for convenience of presentation.)

**The planing machine.** As shaping and planing machines do their work by a series of intermittent cutting strokes, their feed is not a continuous motion as on most other machines, but is arranged to take place during the return stroke of the tool or work table. A diagrammatic representation of a planer feed mechanism is shown at Fig. 67. The vertical splined shaft is driven from the feed motor and timing gear, which synchronises its motion with that of the machine table. The worm drive A imparts motion to the shaft B and gear C, mounted on a clutch which slips in the event of an overload on the mechanism. The pinions D revolve freely as a unit on the vertical feed shaft, the LH pinion being permanently in mesh with C. Motion is given to the vertical feed shaft by the dog clutch E. The cross feed may be engaged by sliding the pinions F so that either the LH pinion meshes with C (forward), or the RH pinion meshes with the RH pinion of D (reverse). The vertical feed is conveyed to the slide through the bevels H to the screw K, revolving in the fixed nut L. Cross feed is obtained through the shaft M engaging with the nut N on the slide.

For clearness, it has only been possible to show the bare essentials of this mechanism, and a number of the refinements have of necessity been omitted. For example, the drive is completely enclosed in a casing with projecting controls for operating the clutch E and sliding pinions F. The arrangement gives a range of 15 feeds varying from $0 \cdot 31$ mm to 20 mm and a quick power traverse.

**Feed values.** For cylindrical work the feed is usually expressed in millimetres per revolution of the work or tool. Thus, if in turning, the tool was advancing $0 \cdot 8$ mm per revolution of the work, it would be equivalent to the turning of a fine thread of $0 \cdot 8$ mm pitch and in 100 revolutions the tool would advance 80 millimetres. Shaping and planing feeds are stated as the distance the tool is fed for each stroke. The feed that may be used for single-point tool cutting depends on (*a*) the smoothness of the finish required and (*b*) the power available, the condition of the machine and its drive. In reference to (*a*), a coarse feed will give an inferior finish to a fine one with the same tool, but with a given feed a tool with a large nose radius will give a better finish than one which is sharper. If too much feed is applied the belt or drive may slip, since the cutting power is proportional to the feed. As a general rule, when roughing work down, the finish is unimportant and the heaviest feed the machine will take should be used. When finishing, the feed should be reduced until it is fine enough to give the class of finish required.

The feeds for drilling are governed by the strength of the drill up to a drill size of about 20 mm and above that on the strength or power of the machine. Drill feeds are stated as hundredths of a millimetre per revolution of the drill, and Table 1 gives a guide as to their values.

Table 1. Twist drill feeds

| Diameter of drill (mm) | 3 | 6 | 8 | 9·5 | 11 | 13 | 16 | 19 | 22–25 | 32 |
|---|---|---|---|---|---|---|---|---|---|---|
| Feed $\frac{1}{100}$ mm/revolution | 7 | 12 | 14 | 16 | 19 | 22 | 25 | 28 | 32 | 38 |

Milling feeds may be expressed as (a) millimetres per minute or (b) millimetres per revolution of the cutter. If a milling machine is of the all-gear drive variety, the pulley runs at a constant speed, and if the feed is driven from this constant speed shaft it may be expressed as millimetres per minute since changing the spindle speed through the gears leaves the feed unaffected. If, on the other hand, the feed mechanism is driven from the main spindle, an alteration in the spindle speed will alter the feed so that it cannot be expressed as millimetres per minute, but must be expressed as millimetres per revolution of cutter. Actually, except from the point of view of rate fixing, it is better to have milling feeds expressed as millimetres per revolution of the cutter because we are better able to judge the conditions under which a cutter is operating. For example, if we are using a cutter with 12 teeth at a feed of 100 mm per minute we know nothing about cutting conditions unless we know the speed. If, however, we know that the above cutter is operating at a feed 0·75 mm per revolution we are able to see at once that each tooth is taking $0·75 \div 12 = 0·063$ mm of cut.

An additional advantage of a feed which varies with the spindle speed is that a raising or lowering of the speed to compensate for easy or difficult cutting conditions is accompanied by a proportionate change in the feed; on a constant feed machine no such change would occur unless adjusted by the operator.

The remarks we made above regarding choice of feed apply in equal measure to milling, with the additional precautionary note that unless the cutter is well supplied with cooling liquid care should be exercised both with feed and speed. Milling cutters cannot be taken out and reground as easily as single-point tools; they are often expensive to replace and their welfare during cutting is not as easily observed as is the case of most other cutting operations.

## Lubrication of the cutting point

On machine tools where the cutting is heavy and continuous, and particularly on milling, drilling and grinding machines, it is essential that there are provisions for delivering an adequate supply of lubricating fluid to the cutting point. The use of a drip-can which delivers intermittent drops, or a thin stream running down a wire, is little more than a joke and has a negligible effect on the real function of cooling and conserving the edge of the cutter. Liquid applied with a brush is even worse, as the tendency there is to add fluid when the cutter seems to be getting hot with the consequent risk of surface cracking, etc. A properly designed lubrication system should include a pump for delivering the cutting fluid which is carried

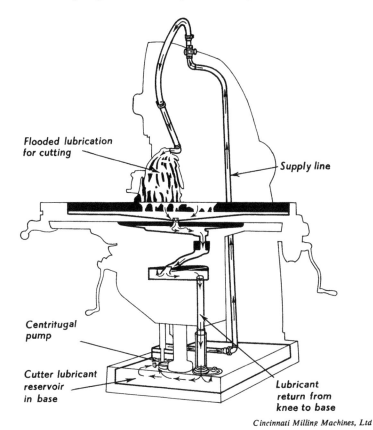

Flooded lubrication for cutting

Supply line

Centrifugal pump

Cutter lubricant reservoir in base

Lubricant return from knee to base

Cincinnati Milling Machines, Ltd

Fig. 68 System for supplying cutting lubricant on milling machine

through piping to a suitable nozzle directed on to the cutting operation. A sump or tank is necessary to carry a working supply of liquid and the pump draws from this, the fluid being caught in the tray or table of the machine after delivery, and drained back to the sump. On most of the modern machines where lubrication is necessary a pump and provision for catching, straining and draining the fluid are incorporated in the design and one such system is shown at Fig. 68. Many machines of older design are without such a system, and where a number of them are arranged in a group it is possible to supply the group from a single pump by arranging a tank overhead, another in the floor and a pump to raise the fluid from the lower to the higher level. A pipe line can be erected to deliver from the overhead supply, with a tap for each machine. The installation is completed by arranging for the delivery to each machine to be caught and drained back into the lower tank.

**Rigidity of machines**

The stiffness and rigidity of machine tools are important factors in the success or otherwise with which they perform their function. If, when under the load of the cut, an undue amount of distortion or deflection takes place in the several portions of the machine, the accuracy of the work will suffer, perhaps to the extent of rendering it unfit for service. A machine which is not solid or rigid enough to absorb and damp out the vibrations promoted by the cutting operation, but allows them to be transmitted to the work or cutter, will produce work of a poor finish, and the vibrations (or chattering) may be serious enough to prevent the operation from proceeding except under very reduced conditions of speed and efficiency. A considerable proportion of the developments and improvements in design that have taken place during the past 30 to 40 years have been with a view to increasing the strength and rigidity of machine tools and, if it is possible for him to do so, the reader should study and compare the designs of machines of similar type and capacity: (a) in the modern version and (b) the design of 40 years ago. An additional factor which has necessitated increased rigidity has been the increase in speed and power made possible by the development of high efficiency cutting alloys, the use of which to their full capacity demands the highest possible properties of stiffness and rigidity in machine tools.

As compared with past designs, modern machines have larger and more robust spindles, increased area of bearing surface on slides, better designed and more rigid tool supporting methods, larger bearings, wider pulleys,

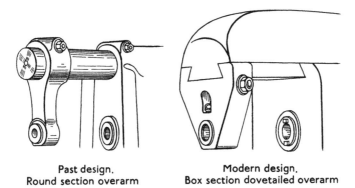

Past design.
Round section overarm

Modern design.
Box section dovetailed overarm

(*a*) Milling machine overarm steady

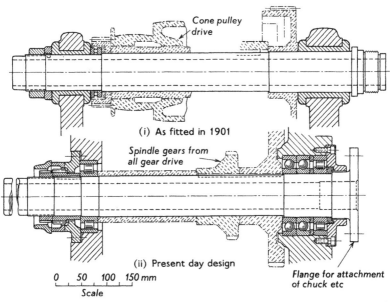

Cone pulley
drive

(i) As fitted in 1901

Spindle gears from
all gear drive

(ii) Present day design

0   50   100   150 mm
Scale

Flange for attachment
of chuck etc

*Alfred Herbert Ltd*

(*b*) Showing development in design of capstan lathe spindle and bearings (spindles
are from same capacity machine and drawn to same scale)

Fig. 69 Showing developments that have taken place in machine-tool design

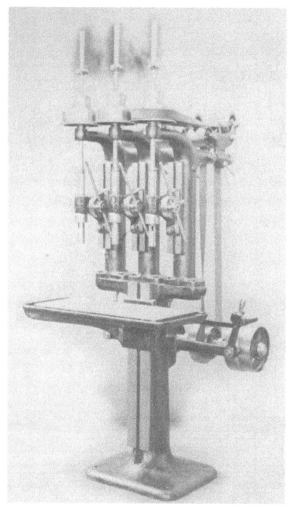

*Fredk Pollard & Co, Ltd*

(c) Design about 1915 (see also Fig. 208)

Fig. 69 (*continued*)

larger adjusting screws, and so on. In addition, their general quality of construction has been improved by the application of improvements in methods of production such as improvements in finish and accuracy of form, ground tooth gears, ground bed surfaces, the use of higher quality alloy steels, cast irons and so on.

In spite of the progress that has been made, however, there are still machines being turned out which are deficient in some of the qualities which promote long life and freedom from wear, absence from deflection, vibration, etc., and the reader should note the good and bad points of various machines so that he may assess the value of different designs. To guide the reader, Fig. 69 shows examples of past and modern practice in the design of various machine-tool details. At (a) is shown the change that has taken place in the design of the arbor support on a milling machine on which the reader will realise the increased stiffness given by the box section overarm and the generous bracket which it carries. The diagram at (b) shows the spindle fitted to the Herbert No. 4 Capstan lathe in 1901, alongside that fitted to the modern design of the same machine. The present spindle is more robust in its proportions and is carried on ball and roller bearings. We should also remember that as well as being more generous in size, modern spindles and other details are produced from higher grade steels than formerly. Advances that have taken place in general design are illustrated by comparing Fig. 69(c) with Fig. 206. Both machines are of identical size and capacity (330 mm), and were made by the same maker. That at (c) was the current design of 1915, whilst Fig. 206 represents modern practice. The reader should note the more generous base and column section, the improved support for the table, the neater and safer driving arrangements given by mounting separate motors for each spindle, and so on.

**Chatter.** In spite of the improvements that have taken place in machine-tool design, however, it is still possible to have chatter and vibration on the tool or cutter and sometimes the remedy for the complaint can be very elusive. Generally, however, the conditions are such that a cure must be found if the job is to be finished satisfactorily, and we give below some of the conditions which promote vibration and chatter.

## 1. Cutting conditions

(a) *Cutting speed too high.* Try reducing the speed of the work or cutter.

(b) *Tool or cutter needs re-grinding.* This can be determined by an inspection of the cutting edge.

(c) *Too much cut or feed.* Try modifications in cut and feed.

(d) *Too little cut or feed.* Tool loiters on its job and chatter commences. When drilling hand material it is better to keep the drill up to its work or it will merely rub and chatter. The same often applies to parting-off in the lathe.

(e) *Tool too small or insufficiently supported.* See that tool is not projecting from its support any more than necessary. Try a tool of larger section if possible.

(f) *Shape, or length of cutting edge.* Tools with a long length of cutting edge such as wide parting and form tools always tend to chatter. Reduce speed or use the tool upside down. Chatter from a round-nosed tool may sometimes be prevented by grinding the tool to a sharper point. Too much cutting clearance may promote chatter by robbing the cutting edge of some of its supporting metal.

## 2. Work conditions

(a) *Work insecurely clamped, badly supported or overhanging.* The remedy for poorly clamped and badly supported work is obvious. Overhanging work may necessitate the rigging up of some form of jack or support to prevent vibration. Vibration on overhanging lathe chuck jobs may often be stopped by merely pressing the tailstock centre against the work.

(b) *Design of work.* Thin work and thin sections on castings tend to vibrate when being machined. Eliminate the drum effect by providing jacks or packing but take care not to distort the job by too much pressure. When cutting with a single-point tool, a sharp-nosed tool will often cut without chatter. Long work on the lathe and grinder should be well supported with the steadies.

(c) *Balance.* High-speed work which is out of balance may give rise to vibrations troublesome during machining. The balance of grinding wheels is important and an unbalanced wheel will cause a form of chatter to take place during grinding.

## 3. Machine conditions

(a) *Slackness in spindle bearings, or end play.* There is not much remedy for slack bearings except by taking up the wear, but a drop of oil may help. The effects of end play may be nullified by an end pressure being applied such as by the tailstock centre on the lathe.

(b) *Slackness in slides.* Take out the slackness by means of the adjusting method provided.

(c) *General 'run down' condition of machine.* Examine the machine periodically for its adjustments and for the tension on the screws and nuts holding it together. A loose nut on the quadrant may allow a vibration to take place throughout the gear train. If the back gear location is at fault the teeth may be allowed to go hard into mesh with the resulting grinding noise, and so on.

### Surface formation and accuracy

As the function of a machine tool is to hold the work and produce the movements necessary for the finishing of its surfaces, we must give some attention to surface formation before we can discuss machine accuracy requirements.

The three surfaces we need to consider are the plane (flat), the cylinder and the cone, as our work is not likely to include any others except on rare occasions. A flat surface is swept out when a plane line moves in its own plane, at the same time as it advances in some direction other than its own. (A plane line is one which lies entirely in one plane.) The two chief varieties

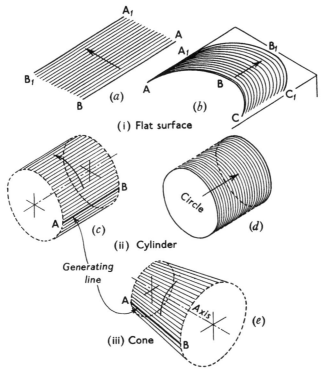

Fig. 70 Fundamental methods of surface formation

of this which concern us are shown at Fig. 70, in which at (a) the line AB advancing to $A_1B_1$ produces the flat surface $AA_1B_1B$, and at (b), the plane circular arc ABC advances to $A_1B_1C_1$ and sweeps out the flat surface bounded by the lines shown.

A cylinder is produced when a straight line rotates about a parallel axis as at Fig. 70(c), or when a circle moves along a straight line perpendicular to its plane (d). When a straight line rotates about an axis in its plane but inclined at an angle, a conical surface is formed (Fig. 70(e)).

### Production of a flat surface

The chief methods of producing a flat surface by machine tools are shown At Fig. 71, and are as follows:

(a) *Shaper, planer and disc-wheel surface grinder.* Tool cuts a straight line and moves over the work in a perpendicular straight line in the same plane.

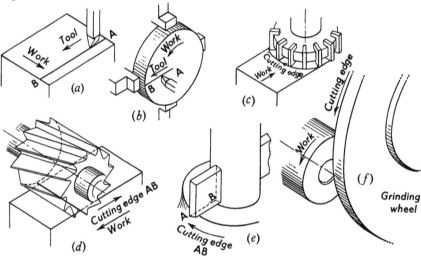

Fig. 71 Methods of producing a flat surface (see text)

(b) *Facing in the lathe.* Tool moves in a straight line and work rotates in the same plane.

(c) *Face milling, end milling, cup-wheel surface grinding.* Cutter edges describe a plane circle which moves over the work in a line in its own plane.

(d) *Slab milling.* Edge of cutter (a straight line) moves in a straight line.

(e) *Spot facing and bar facing* (drilling and boring machines). A straight line AB (cutter edge) forms a flat circle (surface) by rotation in its own plane.

(*f*) *Facing on the cylindrical grinding machine using side of wheel.* Plane circle of cutting edge passes over plane circle formed by rotation of work surface.

### Cylindrical surface

This may be produced as follows (Fig. 72):

(*a*) *Turning, boring, cylindrical grinding.* Combination of a line (tool feed) and a rotation about a parallel axis.

(*b*) *Drilling and reaming.* Same as (*a*) reversed.

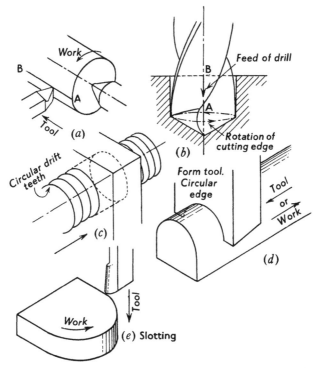

Fig. 72 Methods of producing a cylindrical surface (see text)

(*c*) *Broaching, drifting, etc.* A circle (drift outline) moving along a straight line perpendicular to its plane.

(*d*) *Milling or shaping with semi-circular form tool.* Same as (*c*).

(*e*) *Shaping, planing or slotting.* A line (tool travel) combined with a rotation of tool or work.

### Conical surface (taper)

The formation of taper surfaces is shown at Fig. 73, the available methods being as follows:

(*a*) *Turning, boring and grinding.* A line rotating about an inclined axis in its own plane. (Line may be tool feed as at (*a*) or tool form as at (*b*).)

(*b*) *Reaming.* Same as (*a*).

### Forming and generating

It will be noticed that in some cases the surface form is obtained by a combination of movements on the part of the machine, and depends for its truth on the accuracy with which these movements are performed, whilst in other cases the surface is a copy, or partly a copy of the tool which produces it. For example, in shaping or planing a flat surface, the flat is produced by a combination of two straight-line movements. When a flat face is slab milled its truth in one direction depends on the cutter, of which it is a copy—if the cutter has a hollow in it the surface will have a corresponding bump on it.

Surfaces which are produced by a combination of movements, and depend on the accuracy and relationship of such movements are said to be *generated*, whilst surfaces which are a copy of the tool or cutter producing them are called *formed* surfaces. Some surfaces are a mixture of forming and generating, a screw thread being an interesting example of such. If we cut a vee thread in the lathe with a screw-cutting tool, the thread form is

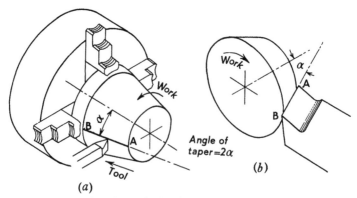

Fig. 73 Methods of producing a taper

a copy of the tool shape and is a formed surface. The screw formation of the thread, however, is brought about by a rotation, and a movement controlled by a combination of gears working in conjunction with the leadscrew, and is generated. As an exercise, the reader might return to the paragraphs dealing with surface formation and decide which methods represent generating and which forming.

## Machine-tool accuracy and alignments

At the quality of the work depends so much on the capacity of the machine to produce various movements and relationships, the accuracy (or alignments) of machines is important.

**The lathe.** The chief requirements of a lathe are that it should turn and bore straight and parallel, and face flat. In order that these conditions must be satisfied the following alignments must be correct.

1. *Turn and bore straight and parallel; work held in chuck.* The tool must move in a straight line which is parallel with the axis of rotation of the work both in plan and elevation. (*a*) If the work axis and tool movement are not parallel in the plan, the work turned will be tapered, being large or small at one end depending on the disposition of the axes (see Fig. 74). If this error is present it is generally due to the headstock not being located on the bed with the spindle axis correctly in line. (*b*) When the work axis and tool movement are out of parallel in the elevation, the edge of the turned surface will not be straight, but hollow. It may be tapered as well.

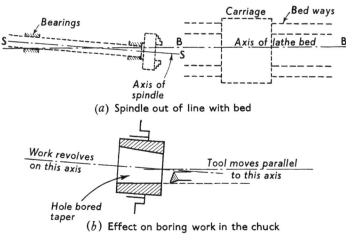

(*a*) Spindle out of line with bed

(*b*) Effect on boring work in the chuck

Fig. 74 Showing cause of lathe boring taper

This fault may be caused through the spindle axis not being parallel with the surface of the bed ways or on old machines, through the ways having worn in one place, the effect of the wear causing the carriage to go up or down-hill slightly at one particularly position. The cause of the hollowness is that the lathe, instead of generating a cylinder is generating a hyperboloid of revolution, the solid swept out when an inclined line rotates about an axis (Fig. 75).

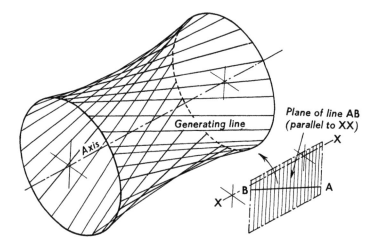

Fig. 75 Hyperboloid of revolution. Hollow solid generated when an inclined line (*AB*) rotates about an axis (*XX*) parallel to its plane

2. *Turn straight and parallel; work on centres.* The axis joining the head and tailstock centres must be parallel with the tool travel (bed axis) in plan and elevation.

3. *Produce a flat face.* In order to face flat, the cross-slide (and tool) must move on a line perpendicular with the axis of revolution of the work. An error, one way or the other, will produce a face either convex or concave. For work held in the chuck, the spindle axis must be in alignment with the bed and the cross-slide perpendicular. If this condition and (*b*) are correct, the machine will also face flat with the work held between centres.

**Alignment tests**

The accuracy in form and relationship of machine-tool elements, such as to render them capable of producing good work, is checked by conducting a set of alignment tests on the machine. Space only allows us to cover a selection of these tests, but the reader may study the subject further by referring to Dr G. Schlesinger's *Inspection Tests on Machine Tools*, in which it is treated fully.

**The Lathe.** The equipment necessary for carrying out the tests for a lathe is as follows:

1. A dial gauge set, the gauge reading to $\frac{1}{100}$ millimetre.

2. A spirit level, having a sensitivity of about $0 \cdot 01$ mm per metre length for each division of deflection.

3. A test mandril, about 25 mm diameter with taper end to fit the machine spindle. Parallel portion about 300 mm long and the whole accurately ground on centres so that parallel and taper portions are concentric.

4. A parallel test bar 25 mm diameter, 300 mm long, ground on its centres.

5. A straightedge, about 300 mm long.

Before making the tests the machine should be levelled up in the longitudinal and cross directions. For clarity and convenience we give the tests in tabular form (see page 110), diagrams being shown at Fig. 76.

**The drilling machine.** The chief requirements for a drilling machine are that the taper hole in the spindle is true and that the machine drills holes perpendicular to the base of the work (i.e. to the table).

Equipment required for tests:

1. A test mandril with taper to fit in the spindle. Parallel portion 100 mm long for machines up to No. 2 Morse and 300 mm long for larger tapers. Shank and taper ground concentric.

2. A dial indicator set.

3. An attachment to fit the taper hole and carry the dial indicator.

4. A block square. This is a frame with square sides and spirit levels incorporated in the base. One vertical side is vee'd to fit against a round column (Fig. 77). In the absence of a block square an ordinary try-square and spirit level will serve.

Before making the tests the base of the machine should be carefully levelled up. (Test numbers in the table (see page 113) refer to Fig. 78.)

| Purpose of Test | Diagram in Fig. 76 | Remarks |
|---|---|---|
| Straightness of bed (longi-tudinal) | 1A Front 1B Back | For vee'd beds, level supported on a block vee'd to suit the ways. |
| Bed level in cross direction | 1C | Take readings at about ten positions along the bed length. |
| Taper hole in spindle true | 1D | Rotate spindle. |
| Spindle axis parallel with bed (vertical plane) | 1D | Rotate mandril to horizontal position of eccentricity if taper hole not true. |
| Spindle axis parallel with bed (horizontal plane) | 1E | Set mandril to vertical position of eccentricity if taper hole not true. |
| Trueness of centre and chuck register | 2A, B and C | Machine shown has chuck flange. When chuck screws on to spindle nose instead of B and C, test the unscrewed portion of the nose. |
| Tailstock barrel parallel with bed (vertical plane) | 3A | Clamp barrel when testing. |
| Tailstock band parallel with bed (horizontal plane) | 3B | — |
| Axis of centres, parallel with bed | 4 | Test in vertical plane only when tailstock has set-over adjustment. |
| Lathe faces flat. . . . | 5 | Face up large blank in chuck. Test this with straightedge and feelers. (Should be flat, to 0·025 mm concave on 300 mm diameter.) |
| Spindle end play . . . . . | — | Set clock plunger against chuck shoulder, and exert end pressure on spindle in both directions. |

**The horizontal milling machine.** In this case we will indicate, in a column of the table on page 115, the effect on the work of an error in the alignment being tested. As before, the machine should be levelled up on its table before testing, and the equipment necessary is as follows:

1. Test mandril about 300 mm long with taper to fit the spindle. Plain portion parallel, about 25 mm diameter and true with taper.

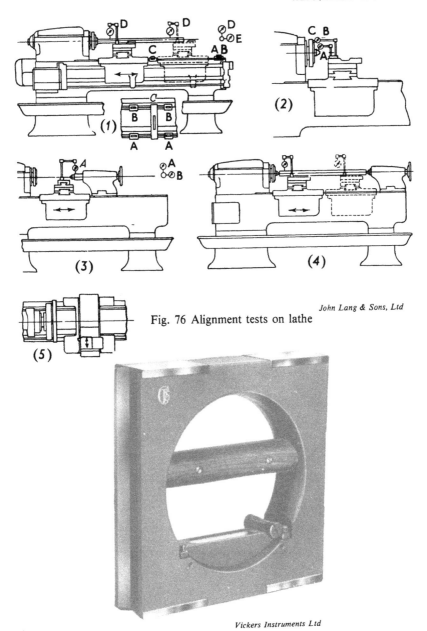

John Lang & Sons, Ltd

Fig. 76 Alignment tests on lathe

Vickers Instruments Ltd

Fig. 77 Block level and square

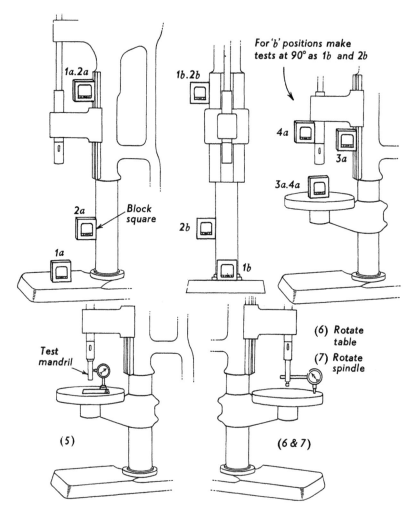

Fig. 78 Alignment tests on drilling machine

| Test No. | Purpose of test | Remarks |
|---|---|---|
| 1a | Column square with base in plane of drill | |
| 1b | do.   perpendicular to 1a | |
| 2a | Drill head guides parallel with column in plane of drill | Use block square or ordinary try-square with spirit level rested on inside of stock |
| 2b | do.   perpendicular to 2a | |
| 3a | Drill head guides square with table | |
| 3b | do.   tested perpendicular to 3a | |
| 4a | Spindle sleeve square with table | |
| 4b | do.   perpendicular to 4a | |
| 5 | Taper hole in spindle running true | Rotate spindle |
| 6 | Work-table true when rotated | Slacken and rotate table |
| 7 | Spindle axis square with table | Rotate dial gauge and take readings at 4 points (90°) |

2. Clock-indicator set.

3. Adaptor to fit arbor, with cross hole and set screw for holding clock-indicator bar perpendicular to spindle axis A. Bar about 375 mm long for holding indicator head (Fig. 79(5a)).

4. 300 mm try-square.

5. Key pieces for centre tee slot and two small angle brackets.

6. Test bar to fit in bore of overarm support bracket.

7. Parallel disc about 150 mm diameter, bored to fit arbor.

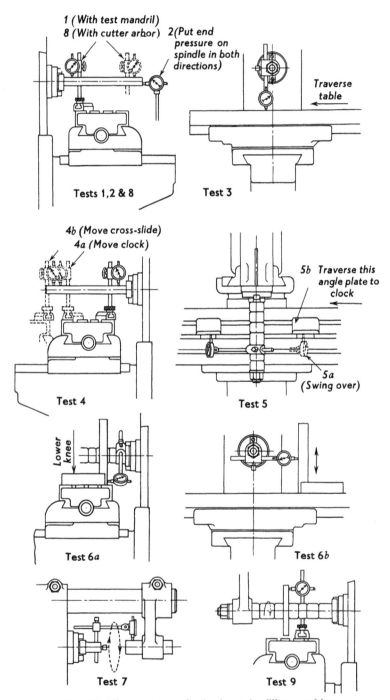

1 (With test mandril)
8 (With cutter arbor) 2(Put end
pressure on
spindle in both
directions)

Traverse
table

Tests 1,2 & 8

Test 3

4b (Move cross-slide)
4a (Move clock)

5b Traverse this
angle plate to
clock

5a
(Swing over)

Test 4

Test 5

Lower
knee

Test 6a

Test 6b

Test 7

Test 9

Fig. 79 Alignment tests for horizontal milling machine

| Test No. | Purpose of test and remarks | Effect of error in alignment |
|---|---|---|
| 1 | Spindle taper true (rotate spindle) | Cut not shared equally between teeth of cutters. |
| 2 | End play in spindle | Cutters will float sideways and cut over, or undersize. Face mills may dig in when leading edges cease to cut. |
| 3 | Parallelism of table in respect to its longitudinal motion. | Milled surface will not be parallel with base. |
| 4a | Table surface parallel with spindle axis. | Vertical face-milled surface not square with base. Horizontal milled face not parallel crossways. |
| 4b | Upper face of cross-slide parallel with spindle axis. | Generally as 4a, also depth of cut will vary when cross-slide is moved. |
| 5a | Centre tee slot square with spindle axis. | Vises and fixtures are lined up from this tee slot. Work milled in them will not have its milled faces correct in relationship with other surfaces. |
| 5b | Centre tee slot parallel with table movement. | |
| 6a | Table square with face of vertical ways. | Vertical milled face will not be square with base. |
| 6b | Table square with sides of vertical ways. | Nothing appreciable for small errors. |
| 7 | Bore of overarm support bracket co-axial with spindle. (Tighten overarm and bracket. Test various distances from spindle.) | End of arbor will be deflected from spindle axis. Work will be out of parallel crossways. Vertical faces may not be square. |
| 8 | Straightness of arbor. (Test with and without overarm support bracket in position.) | Cutters will not run true and cut will not be equally shared by teeth. |
| 9 | Truth of arbor shoulder and collars. (Arbor nut tight.) | Cutters will run out sideways and cut over, or undersize. |

The tests are shown in Fig. 79

# 4 Further work on the lathe

**Promotion of accuracy.** The makers of good lathes are justly proud of their products and the reader may rest assured that any well-known machine, properly used, is capable of turning out good work over a long period. Perhaps, by now, he has had an opportunity of testing the alignments of such a machine and we are sure that he has not found much wrong with them. It is of little use, however, to have accurate machines, unless we plan our work to take full advantage of the possibilities offered. The lathe gives us constancy in the axis of revolution of the work and in the direction of tool travel, a property which enables us to turn and bore any number of diameters concentric, *provided they are all done at the same setting.* The cross-slide, in moving on a line perpendicular to the axis of rotation, ensures that a face, produced at the same setting as a turned diameter, will be square with that surface and that any number of faces will be parallel with one another, and perpendicular to the cylindrical axis. It behoves us, therefore, in planning our work, to take the fullest advantage of these properties, for by so doing we obtain positive accuracy with little trouble. However carefully we set up and locate work for a second setting, there is always the chance that something may move. Our guiding principle should be to machine as many surfaces as possible without moving the job, at the same time giving preference to surfaces which we may know must be correctly related in service. If a second setting is necessary, we should try to plan our work in such a way that from the machining completed in the first operation, the most efficient means of location possible is provided for the subsequent setting of the work. We hope that the following examples may illustrate the points made.

*Example* 1. *To machine the casting shown at Fig.* 80(*a*).

The requirements of the job are that the top diameter of the flange and the two 70 mm recesses shall be co-axial and that the flange faces shall be parallel, and square with the centre line. The inner bore and boss diameter are clearance, and unimportant. As a second setting will be necessary the best locations for it will be on the face and diameter of the flange. This fixes our settings as (1) chuck on boss and (2) strap flange to face plate.

For the first setting the casting should be held on the boss in a 4-jaw chuck and the flange set true, then:

(a) Rough turn face and top diameter of flange, rough bore centre hole to about 53 mm diameter, rough bore recess to about ½ mm less than 70 mm diameter. With a side tool rough face top side of flange to within 9 mm of boss.

(b) Finish, turn and bore the surfaces roughed out in (a). With file and scraper remove all sharp edges.

The operations in (a) and (b) are shown at Fig. 80(b).

The faceplate will be required for the second setting and the top

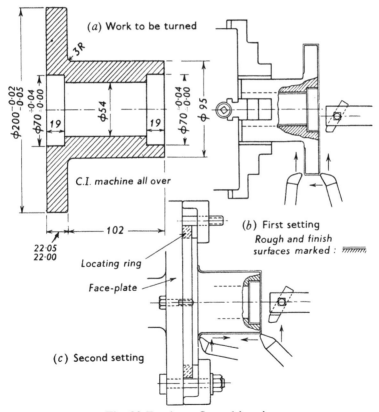

Fig. 80 Turning a flanged housing

diameter of the flange may be located true by one of the following methods:

(i) By holding the component to the faceplate by a long bolt passing right through the lathe spindle, setting the flange diameter true with a clock indicator, clamping the flange to the faceplate with three or four clamps and then taking away the long bolt.

(ii) By attaching a ring to the faceplate with screws from the back, and boring the ring in position to suit the 200 mm diameter flange. The ring should be about 25 mm wide, 13 mm thick and something less than 200 mm in the bore. If much work of this nature is done a stock of various sized rings should be available for this purpose. By providing them with tapped holes about $12M \times 1 \cdot 75$, they can be secured to the faceplate by screwing from the back.

(iii) If four bolt heads or four clamps are attached to the faceplate with their innermost extremities lying approximately on a circle about 197 mm diameter they may be bored out to 200 mm with a boring tool and will then provide a location for centralising the 200 mm flange, which can be strapped to the faceplate with clamps.

Except for light work one of the methods (ii) or (iii) is recommended, as a definite location is provided for the diameter. In method (i) there is no assurance against the work slipping during machining.

When the component has been located and clamped to the faceplate the operations on it are as follows:

(a) Rough turn top diameter of boss to about $\frac{1}{2}$ mm over 95 mm. Complete the facing of the flange roughly, leaving a small amount on for finishing. Rough bore the recess to $\frac{1}{2}$ mm under 70 mm diameter, rough face end of boss.

(b) Finish the facing of the flange and the turning of the boss using a tool with a nose radius of 3 mm. A good way of obtaining a clean corner is first to set the cross-slide indicating sleeve at zero when the tool is positioned to turn the boss to 95 mm diameter. Then face the flange and wind the tool into the corner until zero is reached on the sleeve. Immediately this is reached ease the tool out by operating the carriage and at the same time engage the longitudinal feed to operate *away* from the headstock.

Finish face end of boss and finish bore recess to 70·00 mm diameter, 19 mm deep.

The second setting is shown at Fig. 80(c).

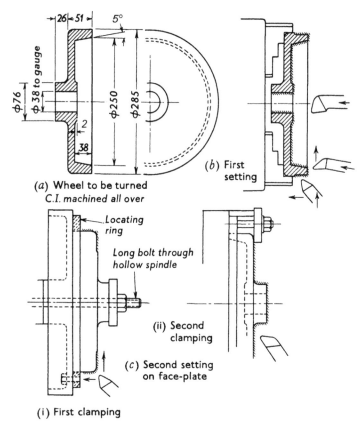

Fig. 81 Operations for turning clutch flywheel

*Example 2. To machine the clutch flywheel shown at Fig. 81(a).*

When it comes for machining, the component will be in the form of a casting with a cored centre hole and with machining allowance all over. The most important requirement is that the tapered portion shall be true with the centre hole and the outside face of the 76 mm boss. The ideal method of holding for the first setting would be to chuck it on the boss, but as this is short, will be tapered and will have a radius where it joins the flange, it is unlikely that a good hold with the chuck jaws would be possible. We therefore propose to chuck the casting as shown at (b) and whilst it is in this position to rough and finish the faces shown shaded. By turning the taper and the bore at the same setting we are assured of their being true with each other and at the same time we are providing a good location for the second setting. The taper must be turned from the compound slide and checked with a vernier protractor.

For the second setting there is a choice of methods available for holding and location.

1. The wheel could be pressed on to a mandril and machined that way.

2. By chucking on the already machined portion of the rim, pressing back to the chuck jaws for side trueness and setting the dimeter with a dial indicator.

3. Machining a locating ring on the faceplate, holding by a long bolt through the spindle whilst the top and about a 25 mm length of the back face are turned, then fitting clamps for the remainder of the facing and turning the boss. The locating ring could of course be dispensed with and the top set true with a clock indicator. This method, with a locating ring, is shown at Fig. 81(c), the two clamping methods being shown at (i) and (ii).

### Taper turning with attachment

Our previous discussion of taper turning was in conjunction with the compound slide, and by setting over the tailstock. Both these methods impose limitations on the scope of the process, and for undertaking a comprehensive range of taper, a taper-turning attachment is necessary. Such an attachment is shown at Fig. 82 where slideway A is supported at the rear of the lathe and is provided with adjustment, indication and locking arrangements for setting and clamping to an angle with the bed of the lathe. B is a sliding fit on A and may be connected to the lathe cross-slide by fitting screw C which clamps it to the arm incorporated with the cross-slide. [It will be seen on this pattern that the cross-slide has upper and lower dovetail ways. The upper ways serve the usual function of carrying the cross-slide, operated by its screw in the usual manner, whilst the lower ways carry the whole of the upper assembly on ways machined across the carriage. When C is in the far hole the lower ways are locked and the cross-slide functions as usual, but when it is moved to the position shown the unit can move across independently of its actuating screw.] To use the attachment, A and its supporting arrangements should be moved to a position along the bed about centrally opposite the taper to be turned. By means of the adjustment and graduations A should now be set to one-half the included angle of the taper, its direction from the central position depending upon whether a male taper is being turned or a hole being bored. Some attachments are graduated in taper per unit length as well as degrees, but care should be exercised as one may never be certain whether the marks refer to the setting of the attachment or to the taper produced on the work which, of course, is double the setting of the attachment. After A has been set as near as possible C is removed from its position of locking the lower slide and fitted in the position shown. The machine is now set for its work and it will be found that when the carriage is moved along, the tool will be led by slide B to follow a path parallel with slideway A of the attachment which is fixed and does not move with the carriage.

On some machines there is no double cross-slide feature such as we have shown, and the arm for connecting B to the cross-slide is fixed to the single, upper cross-slide. This means that before such attachments can be used the cross-slide must be disconnected from its actuating screw to enable it to move across freely and follow the guiding influence of A and B. Under such conditions means are provided for disconnecting the lead-screw nut from the cross-slide so that the latter may move independently of its screw. This is generally done by removing a screw located about in the

*Holbrook Machine Tool Co, Ltd*

Fig. 82 Taper-turning attachment

centre of the upper face of the cross-slide. When the lathe is in this condition the usual cross-slide feed cannot be used, of course, for putting on additional cut, and this has to be operated from the compound slide set perpendicular to the bed. The double-slide feature enabling the usual cross-slide lead-screw to be used whilst the taper attachment is in operation is one of the advantages of this design over those in which the cross nut has to be disconnected.

The operation of taper turning or boring with the attachment differs little from ordinary parallel work. As soon as the taper has been turned down enough for a small length to be tried in the gauge this should be done, as the initial setting of the attachment is sure to need correcting, and with a slow taper we rapidly get down to size (see pp. 342 and 343, Part I).

If a number of similar tapers have to be turned, once the attachment is set correctly there will be uniformity in the work, provided the tailstock is not disturbed and the tool is maintained at the same height (see later).

Occasionally it may be necessary to turn or bore a taper for which neither gauge nor mating part is available for trial. Under such conditions one method is to set the taper attachment as near as possible to half the included angle of the taper, and after taking a trial cut, to check the taper under a sine bar or with a vernier protractor, adjusting the setting and trying again until the taper is correct. Another method is to set the slideway of the attachment to the correct angle. This could probably be done by clamping a sine bar to the slideway and lining the two up. Then working from the edge of the lathe bed, or from an angle plate bolted to the carriage, and lined up from the bed, the sine bar (and slideway) can be set to the angle using a height gauge or other means of obtaining the necessary measurements to the rollers of the sine bar. A similar method of working could be used for setting the compound slide to an exact angle, but for the angle to be accurately transferred to the turned work the tool must be set exactly on the centre as the following analysis shows.

### Effect of tool position in taper turning

In Fig. 83, the radial distance AB in the end view represents one-half the total taper on the job. Thus, if the tool were set on the centre, it would, in travelling along the taper from A to B, move the distance AB out from the centre. If the tool is set a distance $h$ off centre, and starts at D, it would have to move the distance CD to turn the same taper as before. Now CD is greater than AB, so that the tool, if set to move the distance AB but placed on line CD, would turn a taper of less angle. If the reader has diffi-

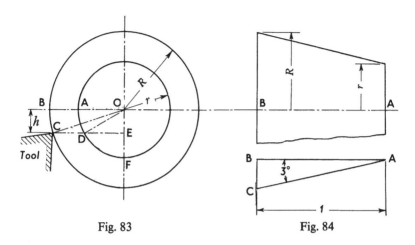

Fig. 83                     Fig. 84

culty in seeing this let him imagine the tool placed at F. If it could be made to cut in this position he will see that it would hardly produce any taper at all! If the taper, the distance $h$ and either $r$ or R are known we can work out the error in a given case as shown in the following example:

*Example* 3. *A taper attachment is set to* 3° *but the tool is* 3 *mm below centre. If the small end of the work is* 25 *mm diameter calculate the actual angle of taper produced.*

The half angle of taper setting being 3°, the tool, in unit length, will move out the distance BC in Fig. 84.

i.e.                    $1 \times \tan 3° = 0 \cdot 0524$ mm

Referring now to Fig. 83 we have that CD $= 0 \cdot 0524$ mm,

$$OE = 3 \text{ mm and } OD = 12 \cdot 5 \text{ mm}$$

$$DE = \sqrt{OD^2 - OE^2} = \sqrt{12 \cdot 5^2 - 3^2} = 12 \cdot 13 \text{ mm}$$

Now        $OC^2 = CE^2 + OE^2$

$$= (CD + DE)^2 + OE^2 = (0 \cdot 0524 + 12 \cdot 13)^2 + 3^2$$
$$= 12 \cdot 18^2 + 3^2 = 157 \cdot 4 \text{ mm}$$

From which OC $= \sqrt{157 \cdot 4} = 12 \cdot 55$ mm

Since OC $=$ OB, this means that

$$AB = 12 \cdot 55 - 12 \cdot 5 \text{ (rad of small end)}$$
$$= 0 \cdot 05 \text{ mm}$$

and the tangent of half the angle turned on the work $= \dfrac{0.05}{1}$ from which the half angle $= 2° 52'$ approx, giving an included angle of $5° 44'$, and an error of $-16'$ due to the incorrect setting of the tool.

### Eccentric turning

Eccentric turning is an operation which, depending on the work, may require the use of the centres, the chuck or the faceplate. If the work is to be accurate it may demand much skill and patience, making the task all the more interesting and worthy of the closest attention. We will discuss one or two examples:

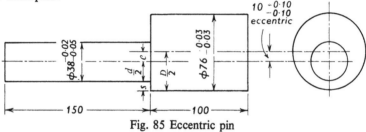

Fig. 85 Eccentric pin

*Example 4. To turn the bar shown at Fig. 85.*

The raw material would probably be in the form of a piece of 80 mm bar about 253 mm long.

1. Hold in the 3 or 4-jaw chuck, set as true as possible, and face and centre one end. Reverse in the chuck, set true, centre the other end and face until bar is about 253 mm long.

2. Ascertain that headstock centre is dead true. Place the bar between centres with a carrier attached and clean up the top diameter parallel. With a knife tool, and half centre in the tailstock skim up the ends to bring the total length to about 251 mm.

3. Put the bar on vee blocks, set the finger of a height gauge to the height of its centre and scribe a horizontal centre line across each end. Turn the bar through 90° and set the previously scribed line vertical, using a square. Increase the height gauge setting by 10 mm and scribe lines at each end for the eccentric centres (Fig. 86(a)).

4. Carefully dot punch and centre drill the ends of the bar, working as closely as possible to the marking out. This centring will be best done on a light sensitive drill.

5. Fit a carrier and place the bar in the lathe, holding on the eccentric

centres. Commence to rough down the reduced portion, keeping plenty of lubricant on the tailstock centre. Test for the lathe turning parallel and set correct if there is any error. As soon as a full diameter has cleaned up, take out and grind the tool to give it a good edge, test for correct tension between centres, then put on a light cut with fine feed and take it along for the full length of the reduced portion.

6. Take out the work and place it between bench test centres (Fig. 144 (a), Part I). If these are not available use the lathe centres, placing a parallel plate across the bed for supporting the height gauge. Set the work *on the centres of the original, short diameter,* and with the eccentric portion at the bottom, test underneath it from end to end with the height gauge. Rotate the bar 90° and make the same test in that position. It is unlikely that the bar will be parallel in both positions (Fig. 86(b)).

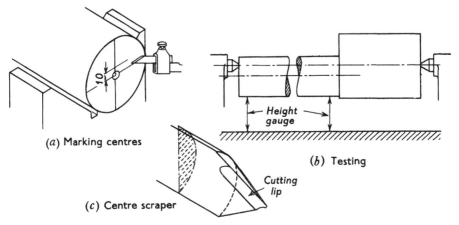

(*a*) Marking centres

Height gauge

(*b*) Testing

Cutting lip

(*c*) Centre scraper

Fig. 86 Operations on turning eccentric pin

7. Assuming that there are errors in parallelism in the above test, the centres of the reduced portion will require scraping to effect correction. Make a note of the amount and direction of the errors to be corrected. Before any scraping is done the 10 mm eccentricity should be checked. To do this rotate the eccentric portion to the bottom and with the height gauge measure the step $S$ (Fig. 85). If $D$ and $d$ are the two diameters and $e$ the eccentricity then $e = \dfrac{D}{2} - \left(\dfrac{d}{2} + S\right)$. This can be determined when the diameters are measured and for guidance in scraping centres its amount at each end should be determined.

8. The centre holes must now be scraped to correct for errors in parallelism of the axes and, if necessary, to bring the 10 mm eccentricity within the limits allowed. A centre scraper may be made from a small half-round file by grinding the end to 60° and coning off the back until it resembles a lathe half centre. A cutting lip should then be ground on the leading flat edge of the 60° angle (Fig. 86(c)). Use this to scrape metal from the countersunk portion of the centre holes on the side which will throw the work over in the direction necessary to correct the faults, and check progress from time to time by testing for parallelism. If there is sufficient metal on the large diameter its centres can be scraped and the parallelism of the reduced diameter checked from them.

9. When the work is deemed to be reasonably correct put the job back in the lathe on the centres that have been scraped, and take a light cut to clean up the portion concerned. Take out and re-test for parallelism and eccentricity, as the centres, bedding in to the scraped centre holes, will probably have caused a slight change to take place. Continue to make any corrections necessary until the axes of the two diameters are parallel in both planes and the eccentricity is within the specified tolerance.

10. Finish turn the 38 mm diameter to length, finish the 76 mm diameter and finish face to length.

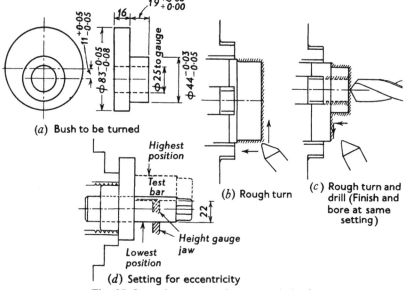

(a) Bush to be turned

(b) Rough turn

(c) Rough turn and drill (Finish and bore at same setting)

(d) Setting for eccentricity

Fig. 87 Operations on turning eccentric bush

*Example 5. To turn the bush shown at Fig. 87(a).*

This would probably be made from a piece of 85 mm diameter bar, 37 mm long.

1. Chuck the material in a 4-jaw chuck, holding on a length of about 6 mm, set reasonably true on diameter and sides, turn top diameter to about 84 mm diameter and clean up the face (Fig. 87(b)).

2. Reverse in chuck, hold on a length of about 12 mm, set 11 mm eccentric using scribing block (reverse chuck jaws if necessary), and set back face parallel with face of chuck (use inside calipers).

3. Clean up front face, rough turn to 45 mm diameter for a length of 19 mm, drill out 24 mm diameter (Fig. 87(c)).

4. Bore hole 25 mm to plug gauge, finish turn to $44 \, {}^{-0 \cdot 03}_{-0 \cdot 05}$ mm and finish two faces to bring the $19 \, {}^{+0 \cdot 05}_{+0 \cdot 00}$ mm correct. (Depth gauge may be used for this.)

5. Clamp the 44 mm diameter portion in the 4-jaw chuck, setting it approximately 11 mm eccentric, and holding with strips of brass to prevent marking its finished surface. Press the component back against the chuck jaws so that the flange will be parallel when faced. When setting eccentric, the 83 mm diameter portion, now to be finished, must be approximately true. The bush may now be set to 11 mm eccentric. Fit a 25 mm test bar in the hole, place a parallel plate over the bed of the lathe and obtain the height gauge. Set the job so that the test plug is at its lowest position and take the reading under it with the height gauge. Turn it round until the plug is at its highest position, take the reading and when the eccentricity is correct the difference in readings should be 22 mm (Fig. 87(d)). Adjust until the setting is correct.

6. Finish turn large diameter to $83 \, {}^{-0 \cdot 05}_{-0 \cdot 08}$ mm and finish face to thickness, checking for parallelism of flange.

7. Remove all sharp edges.

**Aids for eccentric turning.** When a number of similar jobs have to be made, the setting and adjustment necessary is apt to take an undue amount of time. In such cases the use of simple appliances will often help to reduce setting and promote accuracy. Fig. 88 shows an eccentric shaft and one of the end adaptors that could be used for turning on centres. An adaptor is clamped to each end of the shaft, the pair being positioned by the flats which are accurately located relative to the centre holes. The centre holes

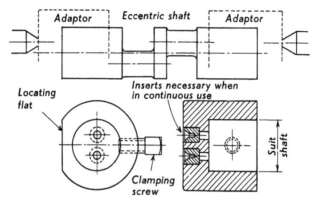

Fig. 88 Turning adaptor

may be formed in the adaptors or a hardened bung with centre hole is driven into a previously jig-bored hole as shown. If there are several throws on the camshaft then a corresponding number of centres, suitably located, may be put in the adaptors. To preserve a chuck setting, a split bush having a bore to suit the component may be set correctly, as shown at Fig. 89. The bush may be opened sufficiently to admit the component by loosening jaw A, which will not upset the setting and which, when tightened again, will clamp the component in the bush.

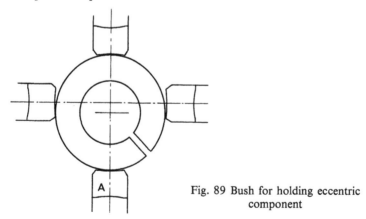

Fig. 89 Bush for holding eccentric component

### Form turning

It is often necessary to turn some unusual form which cannot be obtained by the orthodox use of the machine movements. Common examples occur in turning spherical seated washers, cleaning up large fillets on castings, putting on external radii, forming the teeth on burnishing drifts and broaches, etc. Where the quantity of output justifies the expense, and for everyday jobs such as standard radii, a special form tool may be procurable, and Fig. 90 shows a standard form tool for an external radius, and a special tool. The use of these is straightforward except that when a large tool cutting a good deal of surface is used, the machine must be run very slowly and everything kept tight or chattering and vibration may occur. If chatter does persist, and all other attempts to cure it fail, the tool should be used upside down. This may be done by having the tool at the front and running the machine in reverse, but if the form is unsymmetrical the tool will have to be placed behind the work with the spindle running in the usual direction. For internal radii and other simple forms, it is often possible to grind a form tool by hand with the help of a simple gauge.

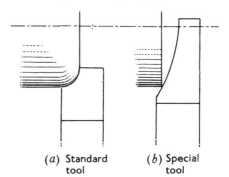

Fig. 90 Radius tools       (a) Standard    (b) Special
                                        tool         tool

When no form tool is available and the form can be finished by filing, a useful method of obtaining it is by hand-working the cross-slide and saddle controls or the cross-slide and compound slide. To do this efficiently some practice, and an intimate knowledge of the machine, is necessary. The form is produced with an ordinary round-nosed tool which, by hand-operating the cross and longitudinal movements in unison, can be caused to follow any desired path, and if a profile gauge is used to check the progress of the work it may be brought close enough to shape to be finished off by filing. In spare moments the reader should practise this on scrap pieces of metal so that he may acquire the feel and technique necessary to carry it out successfully.

**Use of radius bar.** For forming large radii a radius bar gives satisfactory and accurate results. This is a link having a length equal to the radius to be produced, which has one end pivoted to some fixed point on the frame of the machine and the other to the side or end of the cross-slide, depending whether the axis of the radius lies along or across the bed. Fig. 91 shows the use of a radius bar being used for obtaining the concave form of a spherical seating. The bar A is pivoted to strap B fixed across the bed, and at its other end to the cross-slide. The tool must be set so that it is at the centre of the work when the bar axis is parallel with the bed and in use. Cut is put on with the compound slide, the carriage being kept pressed against the bar either by hand or by a cable with a weight hung over the end of the bed. As the cross-slide is operated the radius bar is caused to swing and the pressure kept on the saddle towards it compels the whole of the carriage to follow the longitudinal movement of its end. The tool, therefore, reproduces the circular path swept out by the end of the bar. If the form shown were convex instead of concave the radius bar fixed end would need to be on the chuck side of the tool. To produce radii formed on the diameter of work it would be necessary for the fixed end of the bar to be supported at the front or back of the bed, the bar being parallel with the cross-slide when the tool is in its central position.

**The collet chuck**

For holding and operating on small work up to about 16 mm diameter the large 3 and 4-jaw chucks are not very satisfactory for several reasons.

1. The 3-jaw is seldom true and the 4-jaw requires setting each time.

2. Their jaws mark the surface of bright bars (e.g. silver steel).

3. Tools cannot conveniently be used close up to the chuck.

4. A heavy chuck adds inertia to the spindle and takes away its liveliness.

The collet chuck overcomes these disadvantages and a set of such chucks with the necessary adaptor and drawback arrangement is an acquisition to any small or medium lathe on which there is likely to be a reasonable amount of small work to be done. A section of one of these arrangements in the spindle of a lathe is shown at Fig. 92. A is the chuck which is accommodated in an adaptor B which locates it and is tapered at the end to provide a seating for the tapered chuck end. The drawback tube C passes through the hollow spindle and screws on to the end of the chuck having a nut D fixed to the end which protrudes from the back end of the spindle. To lock the collet the nut is screwed up by means of the tommy-bar holes. This draws the collet into the adaptor and the taper portion, together with the sawcuts, causes the end of the collet to close in

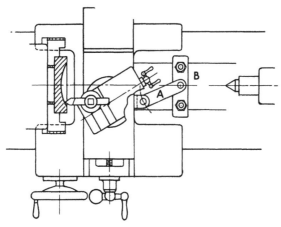

Fig. 91 Spherical turning with radius bar

and grip the work. Sometimes the end of the tube is fitted with a hand-wheel, in which case the tube itself is rotated, drawing in the collet by its screwed connection thereto. The collets are made of carbon steel which can be hardened and spring tempered so that they retain their property of self-opening when the tube is released. The bore of the collet is made to suit the particular size of bar for which it is used, so that to equip a machine for

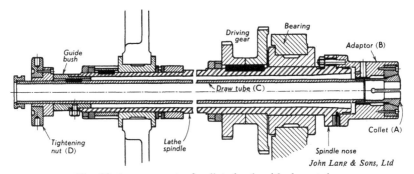

John Lang & Sons, Ltd

Fig. 92 Arrangement of collet chuck with draw tube

a range of work a set of collets will be necessary, having sizes to cover the range. The nature of its location ensures that the work held in a collet shall always run dead true, and as it is supported on a complete circle the arrangement is rigid. The drawback tube being hollow permits bars to be fed through the spindle and this allows small screws, pins, etc., to be turned on the end of the bar and then parted off. Collets may be made with bores square, hexagonal or of shapes to suit other bars of special section.

**Multi-start threads**

The difference between an ordinary single thread and multi-start one is shown at Fig. 93. The 3-start thread at (b) does not look much different from the single thread, but if one thread is followed round it will be seen

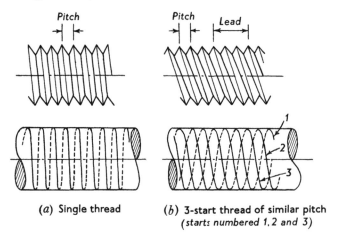

(a) Single thread

(b) 3-start thread of similar pitch
(starts numbered 1, 2 and 3)

Fig. 93 Single and 3-start threads

that there are two more fitted in between, and in one complete turn round the bar this thread advances *three times* as far as the distance between one thread and the next. This distance that any one thread advances along the bar whilst it makes one complete turn is called the *lead,* as distinct from the *pitch* which is the distance between one thread and the next. The different threads are called *starts*. We see, therefore, that in a multi-start with $n$ starts the relationship is:

$$\text{Lead} = n \times \text{Pitch}$$

In a single-start thread, pitch and lead are the same. The size of the thread (height, thickness, etc.) depends on the pitch, whilst the rate at which the tool must travel to cut it (i.e. screwcutting wheels) is fixed by the lead. Hence when determining change wheel ratios the *lead* of the thread must always be used. (Note: lathe lead-screws are always single start.) In addition to this difference in procedure when dealing with threads with two or more starts there are two other points to which attention must be given.

**Locating the starts.** The first, and most important, point is the question of locating the tool in the correct position for cutting the second start after the first one has been cut. If we imagine a four-start thread of 5 mm pitch (20 mm lead) is being cut, when one thread of the 20 mm lead has been put on the shaft there will still be room for three more threads in between and the space will be filled up as the threads are completed. There are various ways in which this can be done: one involves moving the tool, another involves moving the work. Since the distance between one thread and the next is equal to the pitch, if after one thread has been cut, the tool is moved along a distance equal to the pitch it will be in the correct position for cutting the second start. In the case we are considering the tool must be moved 5 mm to the left, and this can be done with the compound slide. The main precaution is to ensure that all the backlash is taken up on the compound slide before commencing the first thread. It is most convenient, of course, to have the compound slide set longitudinally for this movement, but if it is set at some other angle we can still move it the correct amount

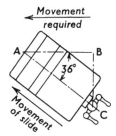

Fig. 94

by allowing for the cosine of the angle, e.g. in Fig. 94 let AC represent the axis of the slide, set to an angle of 36° with the lathe axis. We have to find what length along AC is equivalent to 5 mm along AB, and since $AC = \dfrac{AB}{\cos 36°}$ our movement must be $\dfrac{5}{0\cdot809} = 6\cdot18$ mm along the direction in which the slide is inclined.

If the lathe has no compound slide, or if the one fitted has no graduated sleeve, the second method above may be the more expedient one to use. The starts of a thread are equally spaced round its circumference, so that if, after one thread has been cut, the work can be accurately indexed round by the proportion of its circumference occupied by each start, it will be ready for cutting the next thread. The most convenient way of doing this is to arrange that the gear on the stud has a number of teeth divisible by the number of starts in the thread. When one thread has been cut, carefully drop the bottom quadrant so that the stud gear is freed from mesh and turn the spindle until the requisite number of teeth on the stud gear have been indexed. Then couple up the drive again. In the case we are considering, as there are four starts we might arrange for a 40-tooth gear to be put on the stud, and after cutting each thread index this round 10 teeth. Another and easier method of indexing the work is to have a driving plate with, say, 12 indexing slots equally spaced into any of which the driving pin could be clamped. This would enable the circumference of any job on centres to be divided into 2, 3, 4, 6 or 12 divisions.

**Thread (helix) angle.** The chief noticeable difference between the single- and 3-start threads on Fig. 93 is that the threads of the latter are sloping at a greater angle. The thread angle of a single thread is not generally great

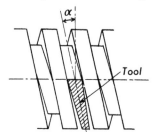

Fig. 95 Tool set at helix
angle of quick thread

enough to cause trouble in cutting, but when the quicker, multi-start threads are being cut, it may be necessary to tilt the tool over to the helix angle to avoid its base fouling the thread. If $\alpha$ is the helix angle then $\tan \alpha = \dfrac{\text{lead}}{\text{circumference}}$ (Part I, p. 285), and from this the angle to tilt the tool can be found. In the thread we are considering, let us imagine it to be a 4-start square thread with top diameter of 38 mm. The mean diameter will then be $38 - 2\frac{1}{2} = 35\frac{1}{2}$ mm, and if $\alpha$ is the mean helix angle

$$\tan \alpha = \frac{\text{lead}}{\text{circumference}} = \frac{20}{35 \cdot 5 \times 3 \cdot 14} = \frac{20}{115} = 0 \cdot 1739$$

from which $\alpha = 9° 52'$ approximately as the angle to set the tool.

This is shown at Fig. 95.

### Lead-screw accuracy

When ordinary common threads are being cut a fraction of $\frac{1}{100}$ mm error in the pitch does not matter much and will not have any noticeable effect on the fit of the screw and nut. When an accurate thread is required, however, such as is necessary on a screw gauge or a tap, accuracy of pitch becomes more important and it may be necessary to have recourse to a machine with a lead-screw of proved accuracy. Unless produced under carefully controlled conditions the pitch of a thread will not be absolutely uniform and accurate but will vary slightly (by variation we only mean in small amounts, probably $\frac{1}{100}$ mm or less). The errors associated with ordinary screws may be *periodic,* or *cumulative* or both. A periodic error is when the pitch increases or decreases by so much in a certain length and then does the reverse in a subsequent length. The error is cumulative when the pitch progressively lengthens or shortens by a certain amount. The lead-screw of the average lathe has both these defects present to some degree and screws cut from such a lead-screw will have the error reproduced on them, together with any faults present in the gear drive to the lead-screw. A chart plotted from the results of a test on a lead-screw is shown at Fig. 96(a), from which the periodic variation will be seen by the rise and fall, and the cumulative error by the gradual upward direction of the whole line. The diagram shows the overall cumulative error to be about $+0 \cdot 11$ mm over the $1 \cdot 1$ metre length, the pitch being too long, and the presence of a local periodic error (shown in inset) occurring at about 12 mm intervals indicates the probability that the screw was cut on a defective lathe with a 6 or 12 mm pitch lead-screw. A screw cut by using the portion EF of this lead-screw would have the periodic variation reproduced, in addition to an increased length of pitch of about $0 \cdot 015$ mm over its 50 mm length.

For precision screwcutting a lead-screw must be used which has been specially made under methods by which the errors are reduced to the smallest possible dimensions. A test on such a screw is shown at Fig. 96(b), where it will be seen that the cumulative error is almost non-existent (within $0 \cdot 010$ mm) and the periodic error very small (less than $0 \cdot 003$ mm). It is usual before reserving a special lead-screw for precision work to have it tested and certified by the National Physical Laboratory, who will issue a Certificate of Examination on which is set out the limits of accuracy of the pitch. A copy of an NPL Certificate is shown at Fig. 96(c).

**Driving for very coarse screws.** When a helix with an exceptionally large lead has to be cut the lead-screw must be geared up by a high ratio and rotated many times as fast as the spindle. Examples of such screws occur in the cutting of oil grooves in bearings where, for example, if an oil groove had to be cut to run from corner to corner of a half bearing 60 mm long, its lead would be 120 mm. For such the gear ratio to a 6 mm pitch lead-screw would be 20 to 1. Under such conditions it may be found impossible

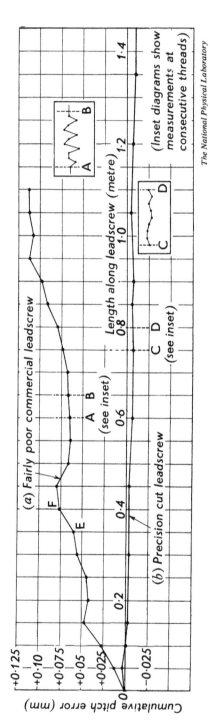

Fig. 96 (a and b) Showing pitch accuracy of ordinary, and precision cut lead-screws

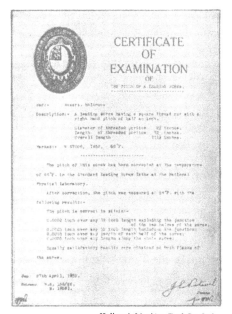

Holbrook Machine Tool Co, Ltd

Fig. 96(c) Lead-screw test certificate

to drive the lathe spindle, the extremely high gear ratio coupled to it causing locking to take place. The way to overcome such a situation is to rotate or drive the lead-screw, allowing it to drive the spindle through the screw-cutting train. The work may then be completed in the usual way.

## Machining a number of similar components

When more than one or two similar articles have to be machined it is often more efficient to vary the method from that which would be adopted for a single component. Instead of machining each job outright before commencing on the next, the machining is divided up into several stages and one stage is performed on every component before going on to the next stage. Working this way, after any particular stage in the process, all the components have advanced that far towards completion.

This method entails more work in changing components into and out of the machine, but reduces greatly the processes of tool changing, setting and trying the size each time, a much longer and more laborious job than changing the work. When the tool has been set for the first component it can always be fed to the same position by making use of the readings on the graduated sleeves and the stops, which are provided for the travel of the carriage and cross-slide. An additional advantage to be gained by working this way is that if the quantity of components necessitates the use of more than one machine, the splitting up of the operations lends itself more readily to the problem of dividing the work between several machines. Older machines and less experienced operators may be given the roughing operations, whilst the finishing may be entrusted to the better machines and operators. The following examples will illustrate how this method of production may be carried out:

*Example 6. To produce one dozen pins shown at Fig. 97(a).*

We will assume that the pins are to be made from 83 mm lengths cut off from 38 mm diameter bar.

In the following sequence of operations, all the components would pass through each operation before proceeding to the next.

1. Hold the material in the 3-jaw chuck with about 22 mm protruding. Face end, centre and reduce to $19\frac{1}{2}$ mm diameter for a length of about $18\frac{1}{2}$ mm.

2. Reverse in the chuck and set about 25 mm protruding. Face end to finished length, centre and reduce to 17 mm for a length of 19 mm (1 and 2 shown at Fig. 97(b)).

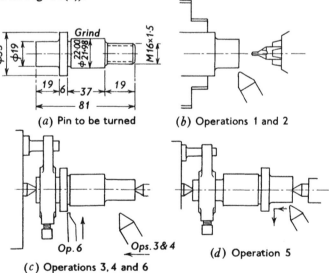

(a) Pin to be turned (b) Operations 1 and 2

(c) Operations 3, 4 and 6 (d) Operation 5

Fig. 97 Settings for turning a pin

3. Remove chuck, fit driving plate and headstock centre. Attach driving dog to the 19 mm diameter end and place between centres. Rough turn top diameter to $35\frac{1}{2}$ mm diameter, reduce to 23 mm diameter for a length of 37 mm less $\frac{1}{2}$ mm (operations 3 and 4 shown at Fig. 97(c)).

4. Finish turn top diameter to 35 mm, finish turn to 22·5 mm diameter $(22+0·5$ mm grinding) for a length of 37 mm, finish turn to 16 mm diameter, 19 mm long, for the thread.

5. Fit dog on 16 mm diameter portion with brass sleeve to prevent marking. Finish turn to 19 mm diameter and face shoulder to give the 19 mm length (use tool with 1 mm rad) (Fig. 97(d)).

6. Face shoulder and undercut (Fig. 97(c)).

7. Screwcut end M16×1·5 p.

8. Chamfer end of screw. Remove all sharp edges.

*Example* 7. *To produce medium quantities of the cover plate (Fig. 98(a))* *on a centre lathe.*

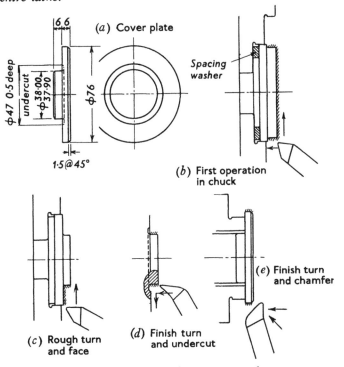

Fig. 98 Turning operations on cover plate

1. Hold in 3-jaw chuck with spacing sleeve to allow 5 mm to be gripped by jaws. Face front and turn top diameter to 76·5 mm up to chuck jaw (Fig. 98(b)).

2. Reverse in chuck. Face to thickness.

3. Same chuck setting. Reduce to 38·5 mm diameter and finish face underside of head to thickness (use a tool with a nose radius of approx. 1 mm (Fig. 98(c)).

4. Same chuck setting. Turn spigot to $\frac{38 \cdot 00}{37 \cdot 90}$ mm diameter and undercut (Fig. 98(d)). Remove sharp edge.

5. Hold lightly on spigot. Turn to 76 mm diameter and chamfer (Fig. 98(e)).

**The square tool post**

From the two examples just discussed it will be realised that if several tools could have been ready set and available for presentation to the job, some of the tool changing could have been prevented. The square tool-post enables up to four tools to be kept in the set position, and is shown at Fig. 60(c). A tool is held on each of the four sides of the post and any of the tools may be rotated and set to the working position. This permits the four tools to be set initially and brought into position when required. On a straightforward job the saving in time is considerable and enables us to combine the division of operations such as we have just discussed, with a set of conditions where working is further facilitated by having several tools set ready for application. The use of this type of tool-post imposes some limitations on the tool position when compared with the usual single tool holders. The four-sided post is generally indexed and clamped into one of the four square positions, whilst on a single toolholder the tool shank may be swivelled to any angle for awkward situations. The limitation is not serious, however, if the work is reasonably straightforward and a good selection of straight and cranked tools is available. As an example of the use of the square tool-post we will indicate its application to the machining of the bush shown at Fig. 99(a).

1ST SETTING

1. Set in chuck holding on a length of about 10 mm.

2. Face end to clean up.

3. Drill centre hole. Put through about an 18 mm drill first followed by one of 35 mm diameter. } 1st tool position (Fig. 99(b)).

4. Rough turn shank to 52 mm diameter.

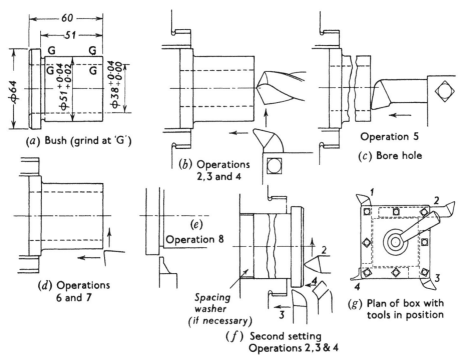

Fig. 99 Turning operations for bush using square tool box

5. Bore hole to 38 mm less 0·15 mm grinding (2nd tool position) (Fig. 99(c)).

6. Finish turn shank to 51·5 mm diameter (for grinding)

7. Finish face end.

} 3rd tool position (Fig. 99(d)).

8. Finish underside of head and undercut (4th tool position) (Fig. 99(e)). Remove sharp edges. A plan of the box with tools in position is shown at (g).

2ND SETTING (Fig. 99(f)).

1. Chuck on shank.
2. Face end (1st tool position).
3. Turn top diameter of head (2nd tool position).
4. Chamfer head (3rd tool position).

### 'Leaving for grinding'

Many turning jobs have subsequently to be hardened and ground or, in some cases for increased accuracy, to be ground although finished in their soft state. The surfaces to be so treated are usually indicated in some way on the drawing, but if no information is given except 'hardened and ground' the turner will have to find out which surfaces he must leave for grinding; generally, those dimensioned to close tolerances and others he may know to be important. The work and patience of the grinder may be helped by the quality of the work he receives from the turning section, and it must not be assumed because a job has to be ground that anything will be good enough.

### Grinding allowance

The amount of metal to be left on diameters for removal by grinding varies according to the work. On external work of normal dimensions 0·4 mm to 0·75 mm on the diameter should provide sufficient material to clean up, and 0·5 mm is the amount commonly allowed. Long shafts which have to be heat-treated before grinding may become distorted, and this may necessitate a more generous grinding allowance to ensure that all parts of the shaft clean up. In extreme cases of such bending, a straightening operation will be necessary before grinding. When leaving for grinding on parts which have to be case-hardened caution is necessary, as if too much is allowed the grinding operation will remove the whole of the hardened outer skin and the work will be useless. When work has to be ground in the soft state after turning a fine turned finish is not necessary; in fact, the grinding is somewhat assisted by a rough surface to commence with.

Owing to internal grinding being a slower method of removing metal the amount to be allowed in holes is less than on external diameters. According to the size of hole and other conditions, internal allowances may vary from 0·03 mm to 0·35 mm on the diameter, and the reader may take 0·15 mm as a suitable amount on average holes. Usually, for work which must be held on an arbor or mandril for some other operation before hardening, turners work to 'grinding' plug gauges which are smaller than standard by the grinding allowance. This ensures that the work will fit the arbor or mandril used during the subsequent operation.

### Undercutting

When a diameter has to be ground sharp up to a shoulder an undercut must be made so that the grinding wheel may render the diameter parallel

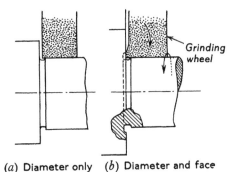

(a) Diameter only   (b) Diameter and face

Fig. 100 Undercutting for grinding

up to the shoulder without touching the latter (Fig. 100(a)). This is because the sharp corner of the wheel, if allowed to run into the corner, would be rapidly worn away. When the shoulder itself must be ground an undercut must be made which encroaches on to the shoulder as well as the diameter. The grinder, with a relieved corner on his wheel, is then able first to finish the diameter, followed by an additional slight lengthwise movement to bring the shoulder against the side of the wheel for facing (Fig. 100(b)). In every case, of course, the depth of the undercut should exceed the grinding allowance left on the work. Blind holes which have to be internally ground should have an undercut at their bottom as wide as it is possible to allow.

### Centre holes

The accuracy with which important work such as mandrils, gauges, test bars, etc., can be ground depends upon the quality of the centre holes. The grinder will probably lap these out with a 60° conical oilstone, but his work will be made easier if they are left in good condition from the turning, and it is easier to smooth them up whilst the metal is unhardened. When otherwise finished, therefore, the centre holes of work should be examined and if rough or torn should be smoothed up with a centre drill. It is well to remember also that the ends of work are seldom touched on the grinding machine, so that poorly finished work, ugly and burred centre holes, etc., will remain as a silent reproach to the inexperienced turner who left them in such a state.

### Boring on the lathe

It is to its versatility that the lathe owes much of its value, particularly in a general engineering shop. In the absence of a milling or boring machine it is possible to carry out a limited range of boring work by clamping the work to the top of the carriage and boring with a bar. The boring bar may be held in the chuck, the tool being accommodated near its end, or it may be of a type supported between the centres, in which case the tool will fit in the bar about midway between its ends. In the latter case, the hole to be bored will have to be large enough initially to allow the bar to be threaded through. In both cases of boring, using this method, the tool has to be moved in the bar when more cut has to be applied. This can be effected by providing some form of screw adjustment for moving the tool small amounts in its longitudinal direction. The feed for boring is obtained, of course, by traversing the carriage in the same way as for ordinary turning. Fastening the work to the top of the carriage may require the construction of adapting arrangements and tackle which will vary according to the lay-out of the carriage top. On some lathes the four corners are faced up and provided with a short tee slot which greatly facilitates the job of securing suitable supporting arrangements for the work to be bored. Other patterns of carriage may require slight modification such as the addition of facings and tapped holes for locating and holding the adapting arrangements, and if the machine is to be used frequently for boring in this way the provision of these will be worth while. Where there is a good height between the top surface of the cross-slide ways and the spindle centre line, a work-table may be made to fit in place of the cross-slide, but this depends on the lathe and upon the type of work to be bored. Fig. 101 shows the set-up for boring

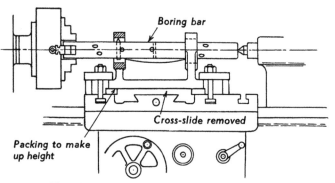

Fig. 101 Bracket being bored on lathe saddle

a bracket in which the hole is initially cored large enough to admit the bar which is supported with one end in the chuck and the other on the tail-stock centre.

**Button boring**

Button boring is a class of work entirely different from that we have just discussed and is used for boring holes, which must be accurately located with regard to their centres, in a flat plate. The process derives its name from the bushes or 'buttons' which are previously fastened to the plate, set in position and serve as media for obtaining the axes of the holes to be bored. Many components of jigs and tools consist of blanks or plates requiring holes accurately located and bored in them, and although the class of work rightly belongs to the jig-boring machine, button boring on the lathe is a good substitute when a jig borer is not available. The buttons used for locating the centres of the holes are plain bushes, hardened and ground to a nominal size (e.g. 8·00 mm. 10·00 mm or 13·00 mm) on their top diameter, their ends being ground square with the diameter. The length of the button is unimportant, a convenient length for a 13 mm diameter button being about 16 mm. (Often one button in a set is made about 4 mm longer than the others to facilitate setting up in the lathe when two buttons are very close together.) The bore also is unimportant, as for the sake of adjusting the position there must be a reasonable clearance between the bore of the bush and the screw which clamps it to the plate. A button and its clamping screw are shown at Fig. 102.

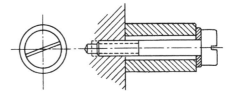

Fig. 102 Boring button

The process of button boring is carried out as follows: The plate to be bored is first marked out reasonably accurately with the positions of the holes. At each hole centre a hole is then drilled and tapped to suit the thread on the button clamping screws. These tapped holes should be as near as possible to the marking out, but slight variations will not matter, as the clearance between the clamping screw and the hole through the button will allow for this. A button is then clamped at each point, and these are set by whatever measuring means are necessary to the exact hole centre positions as given on the drawing. The correct settings being obtained, the button-holding screws are tightened and the plate strapped to the faceplate of a lathe. Dealing with each hole in turn, its button is set true with a clock indicator, the button removed and the hole drilled and bored to size in the usual way. The most important precaution is to ensure that once the buttons have been set correctly they shall not be disturbed during the subsequent operations. At Fig. 103(a) is shown a plate to be bored, the button boring of which could be carried out as follows:

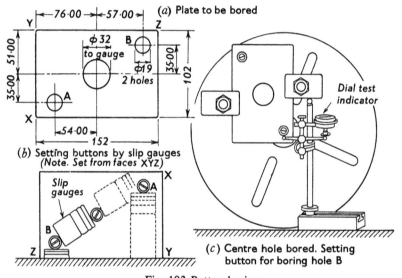

Fig. 103 Button boring

1. Mark out the hole centre lines with a height gauge, the plate being clamped up to an angle plate.

2. Drill and tap the holes for the button-securing screws.

3. Secure and set the buttons in position.

The centre one may first be located by slip gauges working from the edges of the plate, or a height gauge may be used with the plate up against an angle plate. For the other two buttons, one location may be obtained from the edge of the plate and the other from the centre button. If 13 mm buttons are used the distance between the centre one and button A will be $\sqrt{54^2 + 35^2} - 13$ mm $= 64 \cdot 35 - 13 = 51 \cdot 35$ mm and for button B this will be $\sqrt{57^2 + 35^2} - 13$ mm $= 66 \cdot 89 - 13 = 53 \cdot 89$ mm (Fig. 103(b)).

4. Clamp the plate to the faceplate of a lathe and set one of the buttons true with a dial indicator gauge. Remove this button, drill and bore the hole to the plug gauge. Repeat by re-setting to the other buttons and boring these holes. Fig. 103(c) shows the job with the centre hole completed and one of the other buttons being set with the clock.

## Additional methods of support and location

**The running centre** (Fig. 104). When the conditions of a job necessitate the support of the tailstock where, instead of a centre hole, there is a plain cylindrical bore, a large centre mounted on ball bearings may be used. This rotates with the work, and for this reason is called a running centre. These centres are useful for supporting the free ends of hollow cylinders, tubes and other work of a similar nature for turning the top diameter. When using a running centre it should not be assumed that the turned outside

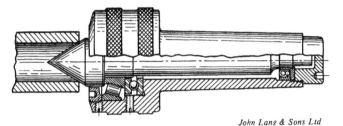

*John Lang & Sons Ltd*

Fig. 104 A running centre

diameter will be true with the bore upon which the support is being taken, unless previous preparations have been made. To ensure concentricity the end of the bore should have been faced true, and to make doubly sure a small chamfer, the same angle as the centre, turned on the mouth of the hole. It is an advantage to use the running centre instead of the usual dead centre for supporting ordinary work with centre holes when the speed is high. The bearing conditions at a centre are so poor that the usual combination cannot stand up for long at high speeds. By using a running centre its perfect point is constantly maintained and the fear of seizing and burning eliminated.

**Expanding peg.** There are many jobs which, after the first setting has been completed in the chuck, require locating and supporting on the bore for facing and finish topping. Examples occur in the case of headed bushes, washers, milling cutters, rollers, etc. A common method of holding such work for the second operation is on a mandril, but this is not always satisfactory for two reasons at least: (1) When a number of components are concerned the use of the mandril entails a walk to the mandril press each time. (2) The mandril is not an ideal method of support and is prone to chattering and vibration. A short expanding peg sometimes offers a better method of holding work of this nature, since the operation of fitting the job is quick and the support is obtained close up to the spindle of the machine. A diagram of such a peg is shown at Fig. 105, in which it will be seen that the taper shank enables it to be fitted to the taper hole in the spindle. The end of the peg has three sawcuts which permit expansion into the bore of the work to take place when the coned screw is tightened.

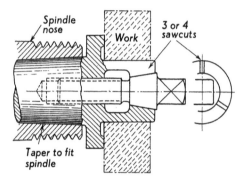

Fig. 105 Expanding peg

**Auxiliary centre holes.** Sometimes the finished shape of a job which has to be turned on centres does not permit a centre hole to be in one or both ends, and if concentricity of its diameters is important a small additional amount of length must be left for an auxiliary centre hole. When the turning has been completed this may then be cut off and the surface cleaned up to its finished shape. Two examples of this are shown at Fig. 106(a), the type of work in each case not permitting of a centre hole in the end. When two or more of such articles have to be turned and their length is not too great

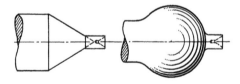

(a) Pip for centre hole

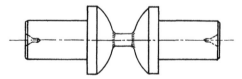

(b) Turning components in pairs to facilitate working
Fig. 106

they may be turned in pairs, being cut apart and finished off when all the turning is otherwise completed (Fig. 106(b)). Sometimes it may happen that a component similar to the ball-ended pin requires turning when in its finished condition. A centre hole must then be obtained by soldering a piece of metal (e.g. brass) to take the place of the pip shown at Fig. 106(a) and then drilling a centre hole in it.

### Slotting in the lathe

The lathe was not designed for use as a slotting machine, but occasionally its use for this operation may provide the way out of a difficulty. Grooves may be cut internally, with a tool after the pattern of a boring tool, or externally, with a tool held on its side in the tool-post. The operation is essentially a hand-operated one, the spindle of the machine being at rest, and the tool moved by working the carriage by hand. Increases of cut are put on by means of the cross-slide. If a number of grooves have to be spaced round the circumference of a job the divisions may be obtained either by the gear indexing method discussed on p. 133 for obtaining the starts of a thread, or by using the lead-screw coupled to the spindle through a suitable gear ratio. Thus, if 20 grooves were required round a job, if the lead-screw were fitted with a 20-tooth gear, driving 100 teeth on the stud (i.e. 1 : 5 ratio), a quarter turn of the lead-screw would index the spindle through $\frac{1}{20}$ of a revolution. Naturally, these methods will not give the extreme accuracy of division necessary for splining, gear cutting, etc., but

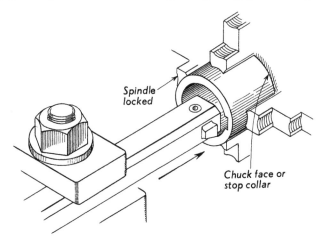

Fig. 107 Slotting a keyway on the lathe

will be good enough for certain jobs, particularly when there is no alternative. A diagram showing an internal keyway being cut with a tool held in a boring bar is shown at Fig. 107.

**Witness marks**

A 'witness' mark is the faint remains of previous machining, of the sawn surface left from cutting off, or of the initial scaly surface left showing after a surface has been machined. Generally this may only be left in one or two places, but its presence may often prove of importance. For example, we

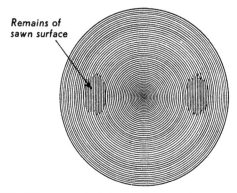

Remains of
sawn surface

Fig. 108 Witness marks left on a bar end faced in the lathe

might be given a piece of material and instructed to turn it up to a certain size, but upon measuring might find it barely large enough for machining to the size required. If we turn it clean and then, after finding it undersize, complain that there was insufficient material on it, we have no proof to support our claim. If, however, we turn it leaving 'witnesses', preferably on opposite sides of a diameter, the remains of the initial surface are there as evidence to verify that the material was not large enough to clean up. The same applies to operations on shaping, milling, grinding, etc., machines (Fig. 108).

# 5 Milling

In the field of machine tool application the milling machine occupies a place opposite to the lathe, the two machines representing the most versatile producers in their respective spheres, i.e. the lathe for cylindrical and the milling machine for plane surfaces. Some millers claim that their machine is more versatile than the lathe, but we must admit that it is easier to do milling on the lathe than it is to turn on the milling machine.

**The horizontal knee type machine**

The most common type of milling machine is the horizontal knee type, so called because of the overhanging 'knee' which slides up and down the front of the machine and carries the cross-slide and table. In our general discussion of machine tools we have already dealt with various features of this machine, but we give a diagram of its complete layout at Fig. 109, the index giving the chief components. Horizontal machines may be of the *plain* or *universal* pattern. The chief obvious difference between the two is that the table of the universal machine is mounted on a turntable and may be swung round in the same way as a machine vise. This feature enables the machine to be used for cutting helices, such as the flutes on a twist drill. In addition, the standard equipment of a universal machine includes a dividing head (see later), and these features, together with other small refinements, make the universal machine the most useful for the range of work likely to occur in the toolroom and general machine shop. The plain miller, although lacking some of these refinements, is often made on more robust lines and is used on production work, where the capacity for heavy cuts is of more value than a wide range of movements.

The arrangement of the knee type machine renders it convenient for operation, as it is open at the front and the work, arbor and controls are easily accessible to the operator. At the same time this feature constitutes a weakness, since the chief force in operation tends to separate the cutter from the work; the arbor deflecting upwards and the knee downwards. The end of the arbor is supported by a bracket carried on the overarm, and to give increased rigidity many machines are supplied with bracing arrange-

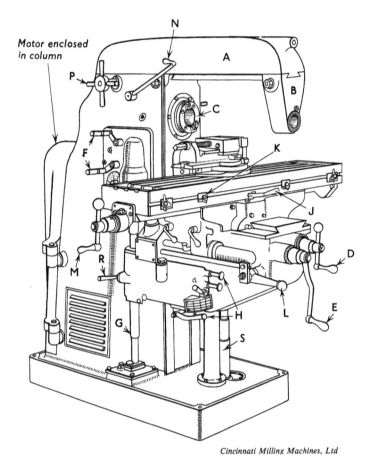

Motor enclosed in column

*Cincinnati Milling Machines, Ltd*

Fig. 109  Horizontal milling machine

*A*, Overarm; *B*, Arbor supporting bracket; *C*, Spindle nose; *D*, Hand cross feed; *E*, Hand vertical feed; *F*, Speed change levers; *G*, Feed driving shaft (enclosed); *H*, Feed change levers; *J*, Table feed actuating lever; *K*, Feed trip; *L*, Rapid power feed control; *M*, Hand table feed; *N*, Starting lever; *P*, Wheel for moving overarm; *R*, Feed reversing lever; *S*, Tube to deliver cutting fluid to reservoir in base.

ments for clamping the knee, arbor and overarm together as shown at Fig. 110. The application of these braces adds to the general stiffness and they are valuable when heavy cuts are being taken, but for light work, especially when frequent changes of cutter are necessary, they are a nuisance, and

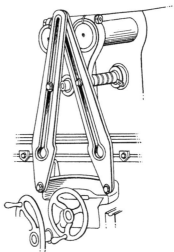

Fig. 110 Milling machine braces

their appearance suggests that they were added to the machine as an after-thought. In a less common design of machine, the bed type, the table is carried on solid metal supported through to the foundation, and the other end of the arbor is likewise very stiffly restrained. Machines of this design naturally lack the versatility of the knee type machine, but their rigidity is a great asset for heavy production work and they are often used for climb milling (see p. 169), using heavy cuts and high feed rates (Fig. 111).

## Accommodation of cutters

Metal is removed on the miller by the application of a revolving cutter in which are incorporated cutting teeth. These may be as solid portions of the cutter or may be arranged as separate blades secured in a body. The cutter may be carried on the arbor or it may be of an end type supported on a taper fitting into the nose of the spindle. When the cutter is carried on the arbor it is provided with a hole in its centre to suit the arbor (usually 27, 32 or 40 mm diameter) and is supported between collars, the assembly being clamped up with a nut. On the smaller machines the cutter is driven by the friction between the cutter-sides and the collars, but the larger types are keyed. The friction drive is an advantage for delicate cutters such as saws, as if the feed is too great or if they dig in they slip and this often prevents damage to the teeth. It is essential that the arbor runs true in all places along its length and that the faces of the collars are flat and parallel. A bent arbor will result in one or two teeth doing all the work, and if the collars are not true their clamping power will be minimised and disc cutters held between them will run out of truth sideways, producing inaccurate shapes and overcutting the work.

End mills, face cutters, drills, etc., fit direct into the spindle with a taper shank. On the smaller types the taper shank is solid with the cutter, fitting the spindle through taper sockets, whilst the larger ones are fitted to a separate stub arbor. Large face mills may be fitted direct to the spindle end without an arbor by being screwed on or bolted to the spigot, according to the design of the spindle nose.

*Marubeni–Iida Co, Ltd*

Fig. 111 Showing construction of bed type milling machine

Within recent years the leading British and American milling machine manufacturers have agreed upon a standardised spindle nose and this, together with the end of the arbor which fits into it, is shown at Fig. 62.

**Milling cutters**

For a milling machine to be fully effective it should be provided with a comprehensive range of cutters, and these should be looked after, not being allowed to depreciate for lack of regular sharpening and maintenance. Milling cutters are made of high-speed steel, hardened and ground. Cutters which fit on an arbor have their bores ground to the standard size of the arbor and their faces ground true with the bore and the teeth, so that, given a true arbor and collars, the cutters will run true. The teeth of end cutters are ground true with the taper. When the cutter is large enough to have separately inserted teeth these are made from high-speed steel and secured into the body of the cutter, which is usually made from a tough, medium carbon steel, unhardened. The cutting edges of the blades are ground up true with the bore or shank of the cutter after the blades have been inserted and secured.

**Cutter teeth**

The teeth of solid cutters may be of the fluted type or machine relieved. Fluted teeth have a shape of the type shown at Fig. 112(a) and are sharpened by grinding the narrow top land of the cutting edge. If the front of the tooth is made radial as at (i), the cutting takes place without top rake, but by deposing the flute as at (ii) a top rate of $\alpha°$ is introduced. Side rake is introduced by cutting the teeth on a helix as at (iii).

When the conditions are such that the tooth must maintain and impart

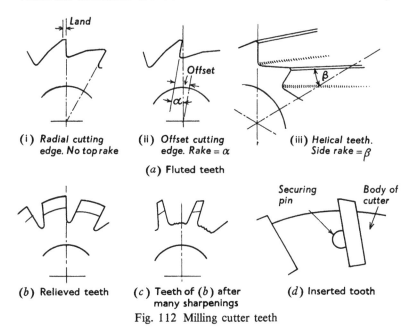

(i) *Radial cutting*    (ii) *Offset cutting*    (iii) *Helical teeth.*
    *edge. No top rake*      *edge. Rake = $\alpha$*        *Side rake = $\beta$*

(*a*) *Fluted teeth*

(*b*) **Relieved teeth**    (*c*) **Teeth of (*b*) after**    (*d*) **Inserted tooth**
                              **many sharpenings**

Fig. 112 Milling cutter teeth

a special and perhaps complicated shape throughout its life, grinding the
top of the tooth would be out of the question, so that for such cases a
machine relieved tooth is employed (Fig. 112(b)). The required shape is
put on the tooth in a relieving lathe with a special form tool, the shaped top
surface of the tooth falling away as shown. For sharpening, the tooth is
now ground on its front surface, so that it maintains its form right down to
the point shown at (c). An inserted tooth is shown at (d).

**Fluted cutters**

*The Cylindrical Cutter* ('slab' cutter, 'helical' cutter, etc.) (Fig. 113(a)).
   This is a solid cutter used for producing flat surfaces, the cross flatness
of which is a copy of the cutter and hence depends on its nearness to a
perfect cylinder. The cutter is sharpened by grinding the tops of its teeth
so that in time the teeth gradually lost their height, and the cutter its
diameter. When the teeth become too low for effective use the flutes must
be ground deeper or the cutter softened and re-cut on the flutes, afterwards

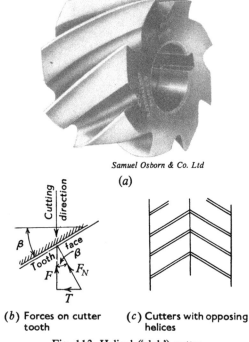

*Samuel Osborn & Co. Ltd*

(*a*)

(*b*) Forces on cutter
tooth

(*c*) Cutters with opposing
helices

Fig. 113  Helical ('slab') cutter

being re-hardened and ground. The teeth on the slab cutter are made helical so that each tooth cuts with a progressive action instead of its whole length coming into action at once, as it would do if it were straight. The effect of the angular tooth also introduces side rake into the cutting action. A disadvantage of the helical tooth is that end-thrust is introduced on the machine spindle, and in Fig. 113(b) if $F$ is the tangential cutting force in the direction of cutting then $F_N$ is the tooth pressure perpendicular to the tooth and $T$ the end-thrust. When two helical cutters are in use at the same time the end-thrusts may be made to cancel out by mounting two cutters having helices of opposite hands (Fig. 113(c)). If the lead and angle of the tooth helix are made suitable it can be arranged that before one tooth goes out of action the next one has come into action; in fact, with a very quick helix, several teeth may be cutting at the same time. The mathematical con-

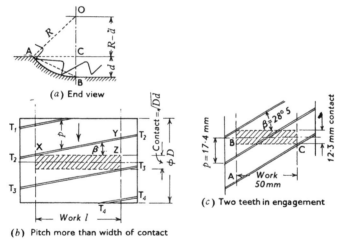

(a) End view

(b) Pitch more than width of contact

(c) Two teeth in engagement

Fig. 114  Cutter tooth engagement

sideration of this is shown at Fig. 114, in which a cutter of radius $R$ (dia $D$), when taking a cut $d$ deep shall just have a tooth coming into action as the preceding tooth goes out of action, cutting taking place over a length $l$ of the cutter. At (a) is shown the end view, where the length of cutter in contact with the work is the arc AB. A simple approximate workable expression for this is given by taking the chord AB which, for the proportions we are considering, differs in length by only a minute amount.

$$AB^2 = AC^2 + CB^2, \text{ and since } AC^2 = OA^2 - OC^2$$
$$AB^2 = OA^2 - OC^2 + CB^2$$
$$= R^2 - (R-d)^2 + d^2 = R^2 - R^2 + 2Rd - d^2 + d^2$$
$$= 2Rd$$

Hence     $$AB = \sqrt{2Rd} = \sqrt{Dd} \quad . \quad . \quad . \quad . \quad . \quad . \quad . \quad . \quad . \quad (1)$$
$$(D = \text{cutter dia}).$$

If the pitch of the teeth is *less* than this there will always be two or more in engagement irrespective of any helical or length considerations.

When the pitch is more than this the conditions are shown at (b), where $T_1$, $T_2$, $T_3$, etc., are the teeth with a helix angle $\beta$ and, as drawn, $T_3$ is just going out with $T_2$ coming into action on the length *l*.

Let *p* be the pitch and *t* the number of teeth in the cutter:

Then     $$p = \frac{\text{Circum.}}{\text{No. of teeth}} = \frac{\pi D}{t} \quad . \quad . \quad . \quad . \quad . \quad . \quad (2)$$

Now     $$\tan \beta = \frac{YZ}{XZ} = \frac{p - \sqrt{Dd}}{l}$$

$$= \frac{\frac{\pi D}{t} - \sqrt{Dd}}{l} = \frac{\pi D - t\sqrt{Dd}}{tl}. \quad . \quad . \quad . \quad . \quad (3)$$

When this has been found the lead of the tooth helix may be calculated

because     $$\tan \beta = \frac{\text{Circum. of cutter}}{\text{Lead of tooth helix}}$$

i.e.     $$\text{Lead of helix} = \frac{\text{Circum. of cutter}}{\tan \beta}$$

We might apply our results to the case of a 76 mm diameter cutter and determine:

1. The teeth necessary for at least two to be in action on a cut 4 mm deep irrespective of helical and length.

2. With this number of teeth the helix necessary for two teeth to be in action on a cut 2 mm deep, 50 mm long.

3. The pitch of the teeth must be less than length AB in Fig. 114(a)

i.e.     $$p = \sqrt{76 \times 4} = \sqrt{304} = 17 \cdot 4 \text{ mm}$$

and from (2)     $$p = \frac{\pi D}{t}$$

$$t = \frac{\pi D}{p} = \frac{76 \times 3 \cdot 14}{17 \cdot 4} = 13 \cdot 7 \qquad \text{say 14 teeth.}$$

4. The conditions for this are shown at Fig. 114(c).

$$\text{Width of engagement} = \sqrt{Dd} = \sqrt{76 \times 2} = 12 \cdot 3 \text{ mm}$$

From Fig. 114(c)  $\tan \beta = \dfrac{\text{AB}}{\text{BC}} = \dfrac{2p - 12 \cdot 3}{50} = \dfrac{34 \cdot 8 - 12 \cdot 3}{50} = 0 \cdot 45$

From which  $\beta = 24° \ 14'$

$$\text{Lead of helix} = \frac{\text{Circum.}}{\tan \beta} = \frac{76 \times 3 \cdot 14}{0 \cdot 45} = \frac{239}{0 \cdot 45} = 530 \text{ mm}$$

It might be interesting to estimate the end-thrust on the above cutter. Assuming 750 W to be absorbed in cutting at a cutting speed of 20 metre/min ($\frac{1}{3}$ m/s)

Tangential force ($F$) (Fig. 113(b)) = $750 \div \frac{1}{3} = 2250$ newtons
     End-thrust ($T$) = $F \tan \beta = 2250 \times 0 \cdot 45 = 1013$ newtons

*The side and face cutter* (Fig. 115(a)). This is a narrow cutter of larger diameter than the slab mill with teeth cut on its sides as well as on its periphery. In the large sizes (over about 200 mm diameter) the teeth are made as separate hardened blades inserted in a tough steel body. The side and face cutter is used in cases where cutting is required on vertical side faces, as well as for cutting slots. It is sharpened by grinding the tops of the teeth, and when this is done the edges of the side teeth are ground slightly saucer-shape so that, instead of cutting on their whole length, they

Samuel Osborn & Co, Ltd

(a) Side and face cutter.                    (b) Slotting.

Fig. 115

operate only on their corners. Often, for cutting grooves and slots, a slotting cutter as shown at (b) is used.

*Angular cutters* (Fig. 116). These are fluted cutters for milling narrow angular surfaces and chamfers, and for fluting other cutters. The *single angle* type (a) has teeth on the angular face and on one side. Cutters of this type are classified as RH or LH. The hand is arrived at from the relation between the slope of the angle and the direction of cutting. If such a cutter when cutting from left to right (i.e. revolving anti-clock) cuts a RH flute (i.e. has its larger diameter nearest the operator), then it is RH. The cutter shown at (a) is RH if it is cutting *away* from the reader. These cutters are classified according to the angle of the groove cut and are commonly made in steps of 5° from 60° to 85°.

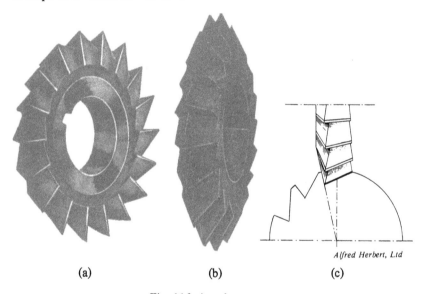

Alfred Herbert, Ltd

(a)                 (b)                 (c)

Fig. 116 Angular cutters

*Double angle cutters* (Fig. 116(b) and (c)) are used for milling angular grooves, and for fluting cutter blanks when the flutes are helical. When used for fluting, as at (c), the one side (that which mills the tooth face) is generally made to an angle of 12°, and when cutting is set radial as shown. The reason for this is that if an ordinary cutter of the type shown at (a) were used, the corners of its vertical cutting face would take away the edge of the tooth being milled as the cutter twists its way along the helical groove. This defect is called interference, and by using a helical slab cutter and an angular cutter as models the reader may observe the effect for himself.

*Saws* (Fig. 117). These are used for cutting narrow slots and for the usual function of sawing. The sides of the saw are ground a few hundredths saucer-shaped, so that the saw body clears as it passes through the slot. Saws are made in thicknesses from about $\frac{1}{2}$ mm, the thin ones being very fragile and necessitating extreme care in their use.

Fig. 117  Slitting saw

*Samuel Osborn & Co, Ltd*

**Machine relieved cutters.** The fluted cutters we have just described are sharpened by grinding the actual cutting edge, the conditions being such that the surface produced by the cutter is a copy of that ground on its teeth at the time of sharpening. For cutter profiles made up of straight

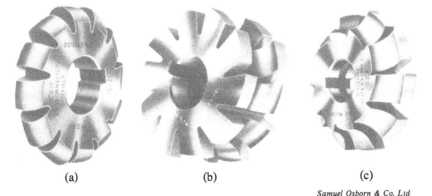

(a)                    (b)                    (c)

*Samuel Osborn & Co, Ltd*

Fig. 118  Form relieved radius cutters

outlines this is easily possible, but curved and special outlines could not be reproduced by such methods of grinding, and machine relieved cutters are used. When making these the blank, previously turned to shape, is gashed and then the tooth form is cut on a relieving lathe.

*Convex and concave cutters* (Fig. 118(a) and (b)). These are used for milling semi-circular grooves and for rounding off the edges of flat sections.

*Radius cutters* (Fig. 118(c)). For milling an external radius subtending a quarter of a circle.

*Fluting cutters* (Fig. 119(a)). The flutes of drills, reamers, taps, etc., are not angular in the same way as those of milling cutters but are of special form. To mill such flutes various shapes of cutters are used.

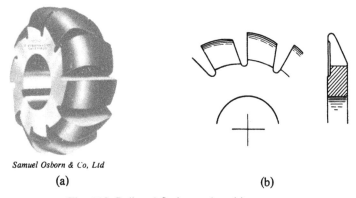

Samuel Osborn & Co, Ltd

(a)          (b)

Fig. 119 Relieved fluting and gashing cutters

*Gashing cutters* (Fig. 119(b)) are used for gashing the blanks of other cutters before machine relieving their teeth.

*Gear cutters* (Fig. 120). These are relieved to the shape of the tooth space and the gear is produced by cutting the tooth spaces, using a dividing head for holding the work and indexing it round for the spacing of the teeth (see later). The milling machine is only used for such work when single or small quantities have to be made, special machines being used for large quantity production.

*Special cutters*. Cutters of the above shapes may be obtained ready made, but for milling special profiles on work, relieved cutters may be made to the required form (Fig. 121).

**Chip breaking.** The edges of large cutters are often nicked for the purpose of preventing a continuous length of chip being formed. This improves and opens up the cutting action, facilitates swarf disposal and economises in power. Such grooves must be staggered on alternate teeth (Fig. 122).

*Alfred Herbert, Ltd*

*Alfred Herbert, Ltd*
Fig. 120  Gear cutter

Fig. 121 Form cutter (double) for cutting teeth of roller chain wheel

**End and stub cutters.** The cutters we have so far described are held on the machine arbor. The equipment of the milling machine, however, includes a selection of cutters which are held in the nose of the spindle without any additional support. Most of these are of the fluted type, although for special forming relieved cutters may be made to this pattern. The vertical milling machine does almost the whole of its work with end and stub cutters.

*The end mill.* End mills are used from about 5 mm diameter upwards. Up to about 38 mm diameter the cutter portion is made integral with the shank,

*Samuel Osborn & Co, Ltd*

Fig. 122  Cutter with nicked teeth

which may be parallel or tapered. Taper shanks are accommodated in the spindle by means of a reducing socket, whilst the parallel shank cutters are held in a special collet. Examples of these cutters, together with a holder for parallel shanks are shown at Fig. 123. The diameter of an end mill has teeth cut on its surface in the same way as a cylindrical cutter, and, in addition, radial teeth are formed on its end. This permits the cutter to be used for milling in the same manner as a slab cutter or by cutting with the end teeth, to mill flat surfaces that way. The cylindrical flutes of end mills may be straight or helical. Straight flutes simplify the manufacture, but when cutting on the diameter have the disadvantage we have already noted in connection with cylindrical cutters. When the flutes are RH helix and the mill is cutting as a cylindrical cutter in a RH direction, the end-thrust tends to pull the cutter from its socket, a happening which can be exasperating,

(*a*) Taper shank.　　　　(*b*) Parallel shank.

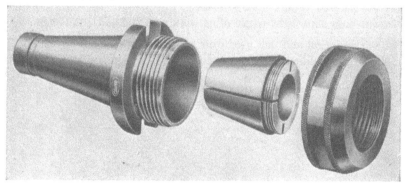

*Alfred Herbert, Ltd*

(*c*) Collet

Fig. 123　End mills

as those who have experienced it know. The hand of the helix affects the cutting rake of the end teeth at the corner—the point where most of the end cutting takes place. If the teeth have a RH helix there will be a positive cutting rake on the end teeth, but if the helix is LH the end teeth will cut with a negative rake. The same applies to the side teeth on a side and face mill with its peripheral teeth helical, and sometimes the top teeth are cut alternate RH and LH helix to give positive rake on the alternate teeth of both sides.

*Shell end mills.* The larger end mills are not integral with their shank, but are made to be screwed, or otherwise fixed on to the end of a stub arbor which has a taper shank to fit the machine. A shell end mill and its arbor are shown at Fig. 124.

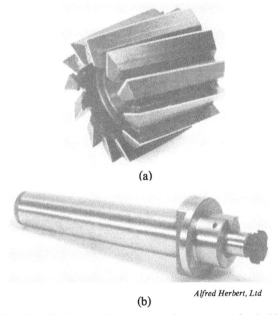

(a)

(b)

*Alfred Herbert, Ltd*

Fig. 124 Shell end mill and type of arbor used for holding

*Other solid end cutters.* The end mill is by far the most common end cutter used for milling, but for special purposes there are various additional types of stub cutters. The *tee slot cutter* (Fig. 125(a)) is useful for milling under-cuts and tee slots, and the *slot drill* (Fig. 125(b)) is for cutting round-ended slots and keyways. Its body is provided with two flutes which terminate with teeth on the end, cutting taking place on the end only. It might be thought that an end mill would provide the most convenient method for cutting slots, but the multiple teeth on small end mills tend to make them clog and drag, so that they wobble and dig into the sides of the slot. The simple two-flute slot drill cuts cleaner and freer with more satisfactory results.

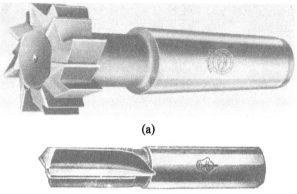

(a)

(b)

*Brooke Tool Manufacturing Co, Ltd*

Fig. 125 Tee slot cutter and slot drill

*The face mill* (Fig. 126). This cutter is used for producing flat surfaces, and for milling represents the most accurate method of production, since the surface is generated and does not depend on the accuracy of cutter grinding as in the case of the slab cutter. Face mills vary in size from about

(*a*) 100 mm to 175 mm diameter. Fixed to a stub arbor.

Fig. 126 Face milling cutters

(*b*) 200 mm to 350 mm diameter. Fixed directly to spindle nose of the machine.

Fig. 126 (*cont.*) Face milling cutters

125 mm diameter upwards and are constructed by inserting blades in a steel body. Cutting rake is imparted by milling the slots which accommodate the blades at a suitable angle. For securing the blades various methods are adopted, two methods for flat blades being shown at Fig. 127.

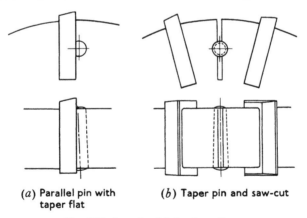

(*a*) Parallel pin with      (*b*) Taper pin and saw-cut
taper flat

Fig. 127 Securing blades in cutters

When the blades are ground for cutting they are provided with cutting clearance and relieved on their fronts, so that cutting takes place at the radius which is ground on the corners (Fig. 128). The smaller sizes of face mills are fitted to their own taper shank arbor which fits into the taper of the spindle, but for large cutters on more recent designs of spindle a spigot is provided to which the cutter body may be secured by screws (see Fig. 126).

### General notes on cutters and milling

Milling cutters are expensive items of equipment and as such deserve to be treated with every care. The average life of a cutter is about 25 re-grinds, and its welfare in between these should be watched carefully, as the symptoms that a milling cutter needs attention are not so obvious as those of a drill or a lathe tool. The weak points of cutters should be studied, and they should be applied in such a manner as to humour their weaknesses as far as possible. For example, a slab mill, when cutting, has its teeth in contact with a long length of the material as compared with a face mill with a reasonably small radius. When machining cast iron upon which there is a crust of intensely hard scale the teeth of the slab mill, being in contact

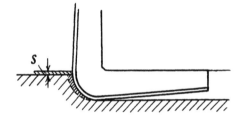

Fig. 128 Face mills blade cutting on radius

with a longer length of this, will have its edge spoiled much sooner than the face mill (compare lengths marked $S$ on Figs. 128 and 130(a)). In this particular case there are two additional factors in favour of the face mill and they are: (1) the edge of the slab mill is much more delicate than that of the other, and (2) when the edge of the slab mill is spoiled its capacity as a producer of flat surfaces has gone, but the face mill can still produce a flat surface whatever may be the condition of its edges. The sharp corner on an end mill or angular cutter is very delicate, and if a sharp corner is required on the work it should first be roughed out with another cutter. When cutting steel a cutting lubricant should be used, particularly when operating with fluted and relieved cutters. The most efficient and satisfactory cutting conditions are obtained with a few, rather than a large, number of teeth in cutter, and the advances that have taken place in cutter design will be seen by comparing the cutters we have illustrated with Fig. 129, which is of a side and face mill, representative of the practice of 125 years ago. Coarse pitch cutters allow better facilities for the tooth to be constructed to the shape giving the maximum efficiency, and the increased space between the teeth allows more room for the chips, preventing their clogging up the cutting action. When rapid metal removal is the primary consideration the cutter should be operated on a heavy feed at a medium cutting speed rather than at a high speed and fine feed. This assumes, of course, that work and cutter are amply supported, otherwise conditions will have to be adjusted to permit the best possible to be done under the prevailing circumstances. When using small end mills and fragile saws the

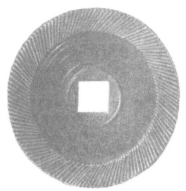

Fig. 129 A side and face cutter
of 125 years ago

*Alfred Herbert, Ltd*

utmost care should be exercised both with speed and feed, as cutters of this type are easily broken. For such cases hand feeding is often preferable, as it is then possible to feel when the cutter is in difficulties and to ease the feed in time to prevent a catastrophe.

## Rotation and feed

When operating a cut with a fluted or machine relieved cutter the normal relation between the directions of rotation and feed are as shown at Fig. 130(a), this sometimes being called 'up cut' milling to distinguish it from the arrangement shown at (b), a less common method of working known as 'down cut' (sometimes called 'climb' milling). The reader is advised not to experiment with down cut milling on the machines he is likely to be using, as the common screw and nut table feeding arrangement is unsuitable, and cannot hold the table back against the tendency for the cutter to overrun itself, climb up and dig in. On heavy cuts, where a considerable arc of the cutter is engaged, down cut working has the advantage that the heaviest cutting pressure is directed downwards and there is no lifting tendency. The method also has certain other advantages, but may only be carried out on

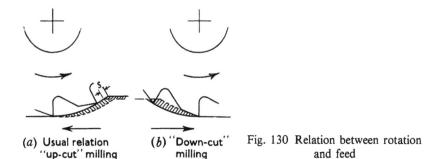

(*a*) Usual relation "up-cut" milling    (*b*) "Down-cut" milling    Fig. 130 Relation between rotation and feed

machines with lead-screw operated tables when they are provided with a backlash eliminator. This ensures that there is never any free movement in the table lead-screw/nut motion, so that the table is always under the control of the nut and cannot be dragged along by the 'digging-in' motion of the cutter teeth. For normal usage the up cut style of working seems unlikely to be supplanted by down cut working.

When surfacing with an end or face mill the relation between the rotation and feed is unimportant except on such occasions as it might be advisable to feed in one particular direction for the sake of some aspect of clamping or supporting, etc.

### Speed and feed

The cutting speed of a milling cutter is the speed of its periphery in metres per minute. The formula, therefore, is the same as that for turning, and if $d$ mm = diameter of cutter; $s$ = cutting speed in metres per minute, and $N$ = spindle speed in rev/min, then

$$N = \frac{1000s}{\pi d}$$

Average cutting speeds for milling

(*High speed steel cutters*)

| Material being cut | Aluminium | Brass | Cast iron | Bronze | Mild steel | High carbon steel | Hard alloy steel |
|---|---|---|---|---|---|---|---|
| Cutting speed metre/min | 16–300 | 45–60 | 20–30 | 25–45 | 20–30 | 15–18 | 9–18 |

**Feed**

The methods used for obtaining the feed on milling machines vary, as also do the terms in which the feeds are expressed. On some machines they are tabulated as millimetres per minute, on others as millimetres per revolution of the cutter, whilst on a large number of the older machines the reader may search in vain for any indication of the values of the feeds. The most equitable method of assessing a suitable feed is by expressing it as so much per tooth of the cutter. This method conveys a truer impression of the actual conditions imposed on the cutter and enables allowances to be made for cutters of a fragile nature, with a small number of teeth, and so on. When the feed per tooth is known the feed per revolution may be found by multiplying by the number of teeth, and the feed per minute is given by multiplying this by the rev/min.

*Example. To find the feed per minute for a 70 mm diameter cutter with 12 teeth operating at a cutting speed of 25 metres per minute and a feed of 0·08 mm per tooth.*

$$\text{If N} = \text{rev/min then N} = \frac{1000s}{\pi d} = \frac{1000 \times 25}{70\pi} = 114 \text{ rev/min}$$

$$\text{Feed per tooth} = 0 \cdot 08 \text{ mm}$$
$$\text{Feed per rev} = 0 \cdot 08 \times 12 = 0 \cdot 96 \text{ mm}$$
$$\text{Feed per min} = 0 \cdot 96 \times 114 = 109 \text{ mm}$$

The table on page 172 will serve as a guide for the feed that may be given to various types of cutters.

**Choice of cutter, set-up, etc.**

In milling, as in other classes of machining, alternative methods will present themselves for the solution of a given job. In deciding between the relative merits of different methods the reader must bear in mind the principles of accuracy we have already discussed alongside the additional peculiarities of the operation of milling. For example, when milling with cutters held by the arbor it is impossible to mill any surface which is to be undercut from the vertical, as the corners of any shaped cutter will foul, remove all the intervening material, and produce a flat, vertical surface. If an angular surface shaped as Fig. 131(a) were required and a cutter having a section similar to (b) were used, when the cutter was rotated and fed across the surface the outermost lower corners of its teeth would cut the job to the dotted line AB. Bearing this in mind, together with our previous remarks

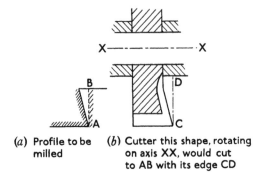

(a) Profile to be milled

(b) Cutter this shape, rotating on axis XX, would cut to AB with its edge CD

Fig. 131 Restriction on milling an undercut

about machining as many surfaces as possible at the same setting, let us consider the milling of the upper and left-hand surfaces of the block shown at Fig. 132(a). For the angular dovetailed slot a stub-mounted angular cutter will have to be employed so that the block might as well be mounted at once with this side facing the spindle of the machine. To complete the whole of this machining at the same setting would then resolve itself into the following operations:

**(1) Working with the arbor**

    (i) Mill top surface—slab mill (Fig. 132(b)).

    (ii) Mill groove—side and face or slot mill (c).

**(2) Working from spindle**

    (iii) Mill side face—shell end mill or face mill (d).

    (iv) Rough out slot—end mill (e).

    (v) Angle out slot—angular cutter (f).

Feeds for milling cutters

| Type of cutter | Cylindrical (slab) mill (up to 30° helix angle) | Slab mill 30°–60° tooth angle | Face mill Shell end mill | End mill | Saw | Slot mill | Form relieved cutters |
|---|---|---|---|---|---|---|---|
| Feed per tooth in $\frac{1}{100}$ mm | 10 to 25 | 8 to 20 | 12 to 50 | 3 to 25 | 5 to 8 | 8 to 13 | 8 to 20 |

When milling hard steel use feeds about $\frac{1}{3}$ to $\frac{1}{2}$ the above

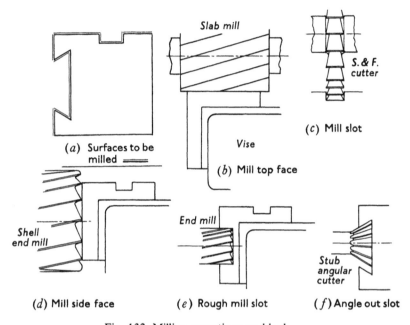

Fig. 132 Milling operations on block

## Care and operation of the machine

We do not propose to explain to the reader how to operate the miller, as someone will show him that. The following notes will help the quality of his work and his quality as a craftsman.

1. Clean down the table and vise (or fixture or dividing head) after each job, and before leaving in the evening. A tee slot scraper (Fig. 133) is invaluable for clearing the slots.

2. Oil or grease the machine regularly with the lubricant recommended.

3. Always use the drawbolt to secure the arbor. Extract the taper by driving it out with a steel bar and not by hammering the drawbolt.

4. Never tighten or loosen the arbor nut without the overarm support in position; you may bend the arbor. Never put the spanner on the arbor nut when the machine is running.

5. Always clamp up the slides not being used to traverse a cut. For this purpose there are generally hand-operated levers on the vertical and cross-slides, and a square or hexagon head set-screw on the table. Check also that the overarm and supporting bracket are tight.

6. Do not drop the cutters on to the table or slides. Their edges and

corners are very hard and sharp. After having had the collars in a pile on the table (probably amongst swarf), wipe their faces before assembling them back again on the arbor.

About
2 mm thick

Fig. 133 Tee slot
scraper

7. Ascertain and remember the value of the graduations on the indicating sleeves of the vertical and cross-slide operating handles. Remember in which direction the table is moved relative to the rotation of its handle. The first two facts will assist in putting on the correct amount of cut; the third may save your putting on cut in a place you did not wish it.

8. Never leave your machine (e.g. to talk to someone) when a traverse is running.

9. The milling machine is classified by the Home Office as a dangerous one. Keep the guard over the cutters, take no liberties and keep hands and tools at a safe distance.

**Examples of plain milling**

*Example* 1. *To machine the component shown at Fig. 134(a).*

1. Set up the machine with a vise on the table and a cylindrical cutter on the arbor.

2. Observing the setting precautions previously discussed, mill the material to 20 mm on the thickness, and $\frac{58 \cdot 00}{57 \cdot 95}$ mm on the width (b).

3. If the cutter being used has nice ground ends and sharp corners on the teeth it can be used for this operation. If not, put on a side and face cutter with sharp corners. Set the solid jaw of the vise parallel with the table. Knock down the job on parallels and mill one step until the thickness is down to 12·97 and width to 50 mm parallel (c).

4. Move the cutter over and mill the other step (d).

5. Turn the job over and mill the clearance on its underside (e).

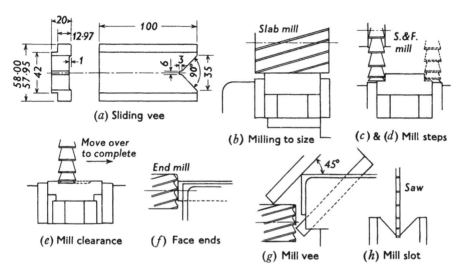

Fig. 134 Milling operations on sliding vee

6. Rotate the vise and set the solid jaw square with the face of the vertical slides. Take out the arbor and fit an end mill about 32 mm diameter. Set the work on parallels and face the ends to length (f).

7. Tilt the job in the vise and set one edge at 45° with a vernier protractor. Mill the vee with the end mill (g). It is apparently important for the axis of the vee to be centrally spaced between the 58 mm sides. Proceed with the milling, working with a rule from the edges until there is about ½ mm left on each face. Then lie a test bar in the vee and check from the edges of the component to the bar. Adjust cut by vertical and cross-slides until vee is central then finish to size by putting on equal amounts of cut by vertical and cross-slides.

(*Note.* This is one of several methods available for milling the vee. It could be done with 90° vee cutter with the block set vertical in the vise, or with S and F Mill and block set at 45°.)

8. Mill the 6 mm × 3 mm slot either with a 6 mm end mill, or with a 6 mm saw. Set the cutter central with the sides of the job before commencing to cut (h).

*Example 2. To mill the keyway and flat on the shaft shown at Fig. 135(a).*

1. Obtain a pair of vee blocks suitable for accommodating the shaft and strap down on the table of the machine. Arrange the positions of the

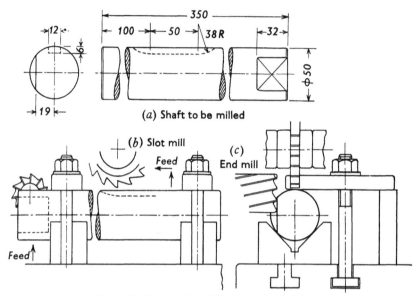

Fig. 135 Setting for milling keyway and flat

blocks and clamps so that both items of machining may be completed without further movement (flat to be milled with an end mill from the spindle). Working from the column face with a height gauge, set the shaft in line with the table.

2. Fit a slot mill about 10 mm wide, 76 mm diameter to the arbor, and using the height gauge as before set its thickness central with the diameter of the shaft. Adjust the cross-slide indicating sleeve to zero, remembering the direction of the backlash.

3. Set the table until the centre of the cutter is 100 mm from the end of the shaft and if you can find the table stop fit this to prevent the table going any further; if not, note the reading or chalk the table lead-screw sleeve.

4. Start the machine, hold the table operating handle with the left hand and with the right hand operate the knee elevating handle until the cutter just scrapes the shaft. Set sleeve to zero. Now carefully raise the knee 6 mm, all the time maintaining a slight tension with the left hand on the table handle, in the direction that the table has to be traversed. As soon as the cutter is sunk in 6 mm, commence to hand operate the table traverse, maintaining a steady movement for a length of 50 mm (if you have another table stop this can be set on the other side to govern your movement, otherwise a rule, or the table lead-screw, must be relied upon. If the lead-

screw is 5 mm pitch it will be 10 turns). As soon as the travel is complete lower the knee (Fig. 135(b)).

5. Micrometer the width of the cutter and add 3 or 4 hundredths according to what you estimate it is overcutting on the width. Subtract from 12 mm and divide by 2. The slot now has to be opened out this amount on both sides. Move the table over by winding the cross-slide the correct amount, sink in to depth and traverse as in 4.

6. Bring the table back to its central position and move over in the other direction taking it 6 or 7 hundredths short to allow for gauging. Sink in a little way and try the 12 mm gauge. Adjust until the gauge is a nice fit then sink in and complete the keyway.

7. Take out the arbor, fit a 40 mm end mill setting and locking the table when the mill is in the correct position to cut the flat 32 mm long. Mill the flat, using the vertical movement of the knee for traversing and finish with the distance from the flat to the opposite diameter measuring 44 mm (Fig. 135 (c)).

*Example* 3. *Operations for machining the casting shown at Fig.* 136*(a).*

1. Set up the casting on its side at the edge of the machine table with about 10 mm overhanging. Pack underneath if necessary to throw base vertical, set rough face vertical and parallel to table and clamp down. Secure stops at ends and behind.

2. Fit a face mill about 180 mm to 200 mm diameter and take a roughing cut over the base, getting well beneath the scale. Continue to machine base until down to size, finishing with a light cut at a fine feed (Fig. 136(b)).

3. Clamp down to base with side overhanging about 10 mm, set rough side in line with table and check that slot also is in line. If not, equalise up the error. Machine the side with the face mill, taking the finishing cut at a fine feed (c).

4. Replace face mill with a 12 mm end mill and rough machine the tenon slot 12 mm wide, and 6 mm deep, with its lower edge about $20\frac{1}{2}$ mm from the base. Raise the knee slightly and take a cut over the lower edge of the slot. Measure from base and finish this edge of the slot 20 mm from base, and on the cut before finishing feed in the end mill to depth. Obtain a 14 mm plug or slip gauge and by machining the upper edge of the slot finish it to width (d).

5. Fit the arbor and put on a large side and face mill. Clamp the job down, setting the machined side parallel with a tee slot and placing the clamps and nuts so that they will not foul the cutter or arbor collars.

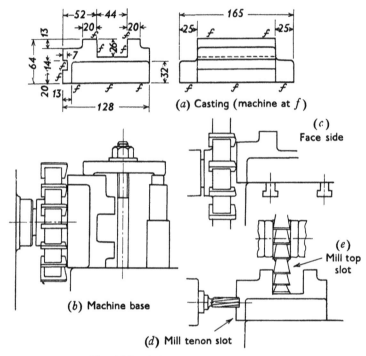

Fig. 136  Milling operations on casting

Rough mill the slot (get well beneath the hard scale) making its side about $52\frac{1}{2}$ mm from the casting side, and its width about 43 mm (e.).

6. Finish the side to 52 mm from the outer face and then the other side until the 44 mm gauge is a nice fit.

7. Machine the top faces of the slot either with the cutter already on the arbor or with a slab mill.

*Example 4. To machine the component shown at Fig. 137(a) from a piece of material 35 mm × 15 mm × 60 mm.*

1. Put on the vise and a slab cutter. Set up and machine the material to thickness and machine one end square.

2. Copper one side by moistening and rubbing with copper sulphate. Mark out and dot punch the profile to be machined. Obtain a 100 mm diameter side and face mill and put it on the arbor.

3. Clamp the work in the vise with a parallel strip underneath so that the marked out radius on one side projects sufficiently above the vise jaw for it to be machined.

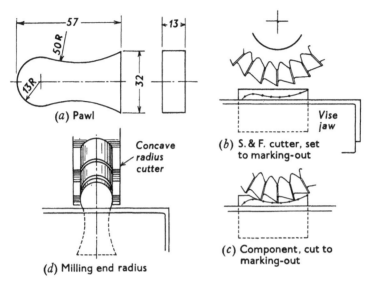

(a) Pawl

(b) S. & F. cutter, set to marking-out

(c) Component, cut to marking-out

(d) Milling end radius

Concave radius cutter

Vise jaw

Fig. 137 Settings for milling pawl

4. Adjust the table movements with the cutter in front of the work so that when sighted directly from the front the outside circle of the cutter teeth coincides with the marked out radius (Fig. 164(b)). Lock the table and lower the knee.

5. Raise the knee until the cutter will take about 1 mm of cut and traverse it *across* using the cross-slide movement. Put on more cut and repeat until the marking out is nearly reached (c). Examine how the cut is related to the marking out, adjust the table if necessary and finish to the marking.

6. Turn the job round and repeat for the opposite side.

7. Take off the S and F cutter and put on a 13 mm rad concave radius, machine relieved cutter. Set the job on its finished end (on a parallel if necessary) leaving the top radius sufficiently projecting to be machined. Turn the vise round and machine the radius on the end (d). To blend the two radii together and finish off the job nicely it will be necessary to touch up with a file where they meet.

### Quantity work on the milling machine. Use of fixtures

An important factor leading to the development and use of the milling machine has been the efficiency with which it lends itself to the quantity machining of small and medium-sized articles consisting of plane surfaces.

In many cases, it was impossible to produce components such as those found in rifles and small arms until the milling machine was designed, although it is to our discredit that it should have been for the production of such articles that the machine owes its development. We have at least advanced to the stage where the principles of mass production, originally evolved for the manufacture of war equipment, are being applied to make things for our own consumption.

The important feature about the milling machine for such work is that the principle it uses of machining its surfaces by means of cutters accurately made to the desired form, enables it to produce large numbers of articles to reasonably close limits of accuracy and *all alike*. This question of similarity is very important where interchangeable quantity production is concerned, as without uniformity in production the whole scheme fails in its application.

### Milling fixtures

The ability to produce large numbers of similar surfaces would be useless without the means of ensuring that the location of such surfaces relative to each other, and to holes and other mating points, were correct. This is obtained by providing a fixture for holding and locating the part to be milled in the correct position for the passage of the cutters. Fixtures are carefully designed and made, so that by their use they make possible a highly efficient and effective means of accurate and uniform production. The design of a fixture incorporates provision for supporting and clamping the work against the forces operating during machining. Means are included for positioning from some previously machined surface, the relationship of which must be held with the machining to be done during the operation in question. Sometimes, if it is an initial operation, the location is taken from certain unmachined surfaces on the casting or forging, choice being made from those which are most influential in the final result. Often, as well as providing location for the component, gauges and setting points are incorporated so that the cutters may be set in the correct position for doing their work. Secondary, but no less important, features included in the construction of milling fixtures are such arrangements as: 1. Facilities for easily renewing surfaces and locating points which become worn in use. 2. Means to prevent the component being put in the wrong way if it is of a shape that such an event is possible. 3. Special equalising arrangements for forgings and castings which may have slight variations on certain of their unmachined surfaces, and so on.

Naturally, the form and appearance of milling fixtures will vary widely since milling embraces an exhaustive variety of components. It is easily possible to have variation between two fixtures each capable of doing the same operation. If only a few hundred parts had to be machined a fairly simple and cheaply made fixture would have to serve (if at all), but if thousands of the same component were under consideration it would be advantageous to construct a more elaborate and permanent method of dealing with the job.

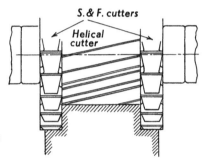

Fig. 138 Milling several surfaces
         simultaneously

**Special cutters**

Often when milling operations are being done on production work it is found possible for several surfaces of the work to be milled at the same time. This often necessitates an assemblage of cutters having sizes bearing special relations, and such cutters have to be made and kept specially for use on that particular operation. An example of this is shown at Fig. 138, in which it will be seen that for the machined width to be correct the centre cutter must be made to a set width, and for the vertical steps the difference between the diameter of the side and face mills and that of the centre cutter is important. The operation of milling several faces at one passage of the cutters is often referred to as *gang milling*, and the cutters as a gang of cutters (Fig. 139). A common form of milling is to finish the surfaces of two opposite faces with a pair of side and face mills spaced apart with collars to give the required distance between the faces. Such an operation is called *straddle milling* (Fig. 140), and this style of milling may often be applied to advantage even when only a small number of components are concerned. To set the cutters quickly to mill any required width necessitates a good selection of collars having various widths (these may be obtained specially for the purpose), and if an adjustable collar is available this enables fine adjustments to be made. A few cigarette packets (for washers)

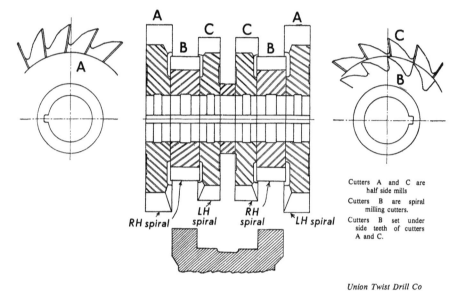

Cutters A and C are half side mills

Cutters B are spiral milling cutters.

Cutters B set under side teeth of cutters A and C.

*Union Twist Drill Co*

Fig. 139  A gang of cutters

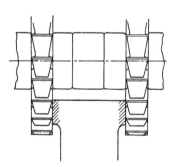

Fig. 140  Straddle milling

or thin steel washers also are useful for setting the cutters to the last few hundredths.

### Examples of fixture work

A simple fixture for straddle milling the width of the large boss on a connecting rod is shown at Fig. 141. The part of the fixture which holds the rod consists of a rectangular block in which is formed an opening (A) shaped to suit the outline of the rod. The opening is extended for some distance beyond the clamping stud (B), so that when a rod is inserted, and the nut tightened, sufficient spring takes place in the metal to allow it to close and clamp the rod. The rod is set by placing the unmachined ends of

Fig. 141 Fixture for straddle milling connecting rod sides

the boss projecting equal amounts on each side from the faces of the fixture. A rod is shown at C.

A fixture for milling the form on turbine diaphragm blades is shown at Fig. 142, together with a finished component in front of the fixture. The

*J. Parkinson & Son*

Fig. 142 Milling fixture

blades are located in an angular slot in the fixture, supported and stopped by an end piece and clamped on reduced portions near their centre. The cutter being used is of the form machine-relieved type.

For holding special nuts to mill a slot a fixture is shown at Fig. 143. This takes two pairs of components and the passage of the two side-and-face mills completes the operation. The fixture concerned provides an interesting example of the equalised clamping of four components by one lever.

## Dividing fixtures

Many milling jobs consist of round or cylindrical-shaped components which, after being milled in one position, must be divided round for a portion of a turn so that the cut may be repeated. Examples of such work occur in the milling of the flats to form square and hexagon heads, saw-cutting across the top face of a castle nut, and so on. The process of indexing round on an ordinary dividing head would be too slow for large quantity production and, moreover, the dividing head is too costly and too much in demand to be tied up on a simple production job. To enable such work to be machined a dividing fixture is used. This has some method of holding the work and incorporates a quickly operated device for turning it through various fractions of a circle. Generally, the divisions are obtained by a plunger which fits into any one of a number of slots or holes equally spaced on a circle or round the edge of a disc. On a general purpose type of fixture the plate might have 12 slots to enable the work circumference to be split into 2, 3, 4, 6 or 12 divisions. An example of a semi-automatic type of dividing fixture being used for a vertical milling operation is shown at Fig. 144. The fixture is indexed by means of a sliding rack which is anchored to a special bracket carried on the column of the machine. This rack projects through the fixture and works in conjunction with gears which in turn withdraw the locating plunger from the dividing plate and revolve the work-holding spindle. This operation takes place on the return stroke of the table and is actuated by hand motion at the completion of each cut. The reader should note the low, squat design of this fixture in order that it may have the maximum rigidity.

## Miscellaneous set-ups

Photographs of miscellaneous jobs being machined on the horizontal miller are shown at Fig. 145. The component shown at (a) is 135 mm long, the machined sides 38 mm deep and the bottom faces about 13 mm wide. The

*Alfred Herbert, Ltd*

Fig. 143 Fixture for slotting special nuts

*Alfred Herbert, Ltd*

Fig. 144 An indexing fixture (see text)

*James Archdale & Co, Ltd*

(a)

*James Archdale & Co, Ltd*

(b)

Fig. 145 Examples of milling operations (see text)

*Cincinnati Milling Machines, Ltd*

(c)

*J. Parkinson & Son*

(d)

Fig. 145 (*cont.*) Examples of milling operations (see text)

cutters are 160 mm diameter, 44 mm wide and on the roughing cut remove 4 mm of metal (cast iron) at a cutting speed of 17 metres/minute and a feed of 38 mm per minute. The finishing cut removes a few hundredths from the bottom faces only. At the setting in Fig. 145(b) the face shown is 400 mm × 220 mm, and is milled with 230 mm face mill at 21 metres per minute and 165 mm per minute feed. The roughing and finishing cuts are 5 mm and 0·25 mm deep, respectively. After this operation and at the same setting the two seatings on the top edge of the face are milled with a shell end mill. The faces being milled at (c) are the slide faces of a machine casting. Note the construction of the fixture holding the work, and the opposing cutter helices to eliminate end-thrust.

Fig. 145(d) shows the body of a face mill being milled with the blade slots. The slots are 13 mm wide, 19 mm deep and are completed in two operations. The first cut round is made 18 mm deep with a 10 mm wide side and face mill at 19 metres per minute, and a feed of 30 mm per minute. The second, finishing cut to the full depth, is made with a 13 mm cutter at 38 mm per minute feed.

*Cincinnati Milling Machines, Ltd*

Fig. 146  Slotting attachment for milling machine

## Slotting on the milling machine

When discussing the slotting machine (later) we shall see how useful this style of machining is for certain classes of work, and that in some cases it is the only alternative to long and laborious hand methods. In many small shops the full-time use of a slotting machine is not justified by the amount of suitable work, and for such situations an attachment by which slotting may be done on the milling machine is available. This is bolted to the vertical face in front of the spindle and is driven by an adaptor which is accommodated in the end of the spindle. Such attachments usually provide for a maximum stroke of about 50 to 75 mm and this is found to be quite sufficient for a large number of keywaying and similar jobs for which the method is so useful. The slotting tools used with such an attachment have nose profiles on the same principle as for orthodox tools, but are made to smaller dimensions. A slotting attachment used in conjunction with a circular table greatly widens the scope and usefulness of a milling machine. A diagram of this attachment is shown at Fig. 146.

# 6 Milling *(continued)*

### The dividing head

An important use of the milling machine is for cutting slots, grooves, teeth, etc., which are to be spaced round the circumference of a cylinder or disc. Examples of such work are given by the teeth of gears, ratchet wheels, milling cutter blanks, reamers, and so on. Milling of this nature necessitates the assistance of some means of holding the work, and rotating it the correct amount for each groove or slot to be cut. The degree of accuracy to which the spacing must be made is important, as much of the work concerned is essentially of a precision character (e.g. gear teeth, splined shafts, cutter teeth, etc.), and demands the highest grade of workmanship for its success. To adapt the milling machine for such purposes a dividing head is used. This is a head which is lined up and bolted to the machine table at the right-hand end and consists essentially of a spindle to which is keyed, inside the head, a 40-tooth wormwheel. Meshing with this wheel is a single-threaded worm, the spindle of which projects from the front of the head and has a crank and handle attached. The head spindle is bored through with a taper hole and is also screwed on its end like a lathe spindle, so that a chuck may be fixed on for holding work not suited to be carried on centres. When

*J. Parkinson & Son*

Fig. 147 Universal dividing head and tailstock (with work steady)

work is held on centres a tailstock is available which is located, for lining-up purposes, in the same tee slot as the head, and bolted down at a suitable position along the table. The head may be rotated so that its spindle is inclined at any angle between the horizontal and the vertical to make it adaptable for conical or end work. In addition, for obtaining small inclinations from the horizontal with work on centres the centre point of the tailstock has a limited up and down adjustment. A diagram of the head with its tailstock is shown at Fig. 147, whilst the mechanism of the head is shown diagrammatically at Fig. 148.

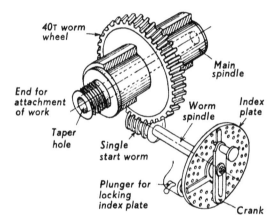

Fig. 148 Diagrammatic sketch showing principle of dividing head

**Index plate**

Since the gear ratio in the head is 40–1, 40 turns of the crank cause the spindle (and the work attached to it) to make 1 complete turn, or 1 turn of the crank rotates the work $\frac{1}{40}$th of a turn. The object of the index plate with its holes is to enable 1 turn of the crank to be further subdivided, as it has to be if divisions other than factors of 40 are required. Normally this plate is locked in position as shown by the plunger in Fig. 148. In the plate are a number of circles of holes, each circle containing a different number, and the spring-pin on the crank can be adjusted to a radius such that it will fit in any desired ring of holes. The two sector arms shown on the front of the index plate at Fig. 148 can be adjusted to include any angle and are for dividing off any desired proportion of a hole circle circumference. The scope of the head (i.e. the number of divisions possible) will depend on the range of hole circles available for enabling subdivisions of worm rotation to be effected, and the provisions made on the leading makes of dividing heads do not leave many gaps in the range of divisions likely to be required.

The Brown and Sharpe head is provided with three index plates having hole circles as follows:

Plate No. 1.   15, 16, 17, 18, 19 and 20 holes.
Plate No. 2.   21, 23, 27, 29, 31 and 33 holes.
Plate No. 3.   37, 39, 41, 43, 47 and 49 holes.

The index plate used on Cincinnati and Parkinson heads is of larger diameter than the Brown and Sharpe plates and is reversible. It is provided with the following hole circles:

On one side:        24, 25, 28, 30, 34, 37, 38, 39, 41, 42 and 43 holes.
On the reverse side: 46, 47, 49, 51, 53, 54, 57, 58, 59, 62 and 66 holes.

With this single plate all divisions up to 60 may be obtained, all even numbers and numbers divisible by 5, up to 120 and most of the commonly required divisions up to 400. For indexing divisions outside these ranges either special plates may be used or a system of differential indexing employed.


**Indexing**

When setting out to index any desired number of divisions the first step is to determine the number of turns of the worm (and crank) necessary to turn the head wormwheel and spindle through the portion of the circle concerned. Since 40 turns of the crank cause 1 turn of the work, if we require $n$ equal divisions of the work each division will be $\frac{1}{n}$th of its circumference and the crank turns will be $\frac{40}{n}$. It will help the reader to remember which way up this fraction should be if he remembers that when *more* than 40 divisions are required the crank will have to rotate *less* than one complete turn, and vice versa. If $n$ is less than 40 the fraction must be divided out into a whole number and a fraction, the latter then being converted to a fraction having a denominator the same as one of the whole circles. Then the numerator will represent how many holes must be indexed in addition to any complete turns obtained from the division. If $n$ is greater than 40 the fraction must be converted to a suitable denominator as above, the indexing then being the number of holes equal to the numerator.

*Example* 1. *To index the following divisions on a Brown and Sharpe head* (*a*) 12, (*b*) 17, (*c*) 25, (*d*) 36, (*e*) 52, (*f*) 86.

12 Divisions

Indexing $= \frac{40}{12} = 3\frac{4}{12} = 3\frac{1}{3} = 3\frac{6}{18}$.

i.e. 3 complete turns and 6 holes in an 18-hole circle.

17 Divisions

Indexing $= \frac{40}{17} = 2\frac{6}{17}$

i.e. 2 complete turns and 6 holes in a 17-hole circle.

25 Divisions

Indexing $= \frac{40}{25} = 1\frac{15}{25} = 1\frac{3}{5} = 1\frac{12}{20}$

i.e. 1 complete turn and 12 holes in a 20-hole circle.

36 Divisions

Indexing $= \frac{40}{36} = 1\frac{4}{36} = 1\frac{1}{9} = 1\frac{3}{27}$

i.e. 1 complete turn and 3 holes in a 27-hole circle.

52 Divisions

Indexing $= \frac{40}{52} = \frac{10}{13} = \frac{30}{39}$, i.e. 30 holes in a 39-hole circle.

86 Divisions

Indexing $= \frac{40}{86} = \frac{20}{43}$, i.e. 20 holes in a 43-hole circle.

**Angular indexing.** Sometimes it is necessary to cut grooves or slots subtending a given angle at the centre of the circle upon which they are spaced. In this case it must be remembered that 1 turn of the crank indexes the work $\frac{1}{40}$th of a turn i.e. $\frac{360}{40} = 9°$ of angle. Hence the indexing necessary to obtain any angle will be

$$\frac{\text{Angle required (degrees)}}{9}, \text{ or, working in minutes, this will be}$$

$$\frac{\text{Angle (minutes)}}{540}$$

*Example* 2. *To index the following angles on a Cincinnati head:* (*a*) 38°, (*b*) 49° 30′, (*c*) 61° 20′, (*d*) 8° 15′, (*e*) 24° 36′.

(*a*) 38°. Indexing $= \frac{38}{9} = 4\frac{2}{9} = 4\frac{12}{54}$

i.e. 4 complete turns and 12 holes in a 54-hole circle.

(*b*) 49° 30′. Indexing $= \frac{49\frac{1}{2}}{9} = 5\frac{1}{2}$  i.e. $5\frac{1}{2}$ turns.

(*c*) 61° 20′. Indexing $= \frac{61° \ 20′}{9} = \frac{61\frac{1}{3}}{9} = 6\frac{7\frac{1}{3}}{9} = 6\frac{22}{27} = 6\frac{44}{54}$

i.e. 6 complete turns and 44 holes in a 54-hole circle.

(*d*) 8° 15′. Indexing $= \frac{8° \ 15′}{9} = \frac{8\frac{1}{4}}{9} = \frac{33}{36} = \frac{11}{12} = \frac{22}{24}$

i.e. 22 holes in a 24-hole circle.

(e) 24° 36'. Indexing $= \dfrac{24° \, 36'}{9} = \dfrac{24\frac{3}{5}}{9} = 2\dfrac{6\frac{3}{5}}{9} = 2\dfrac{33}{45} = 2\dfrac{11}{15} = 2\dfrac{22}{30}$

i.e. 2 complete turns and 22 holes in a 30-hole circle.

**Setting the sector and operating.** When the plunger has been adjusted to fit in the required hole circle the sector must be set to space off any odd holes occurring in the movement. With the plunger in any hole start counting clockwise *from the next hole*, count off the number of holes required and set the sector arms to embrace the plunger, and the last hole

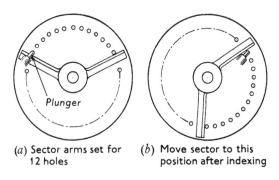

(a) Sector arms set for    (b) Move sector to this
12 holes                         position after indexing

Fig. 149  Operation of dividing head sector arms

counted. This is shown at Fig. 149(a), in which 12 holes are spaced off. When winding the crank to index rotate it clockwise, and after the movement, which will terminate with the plunger in the hole bounded by the *opposite* sector arm from which the start was made, immediately *move the sector round until its other arm is touching the plunger* (Fig. 149(b)). It will then be ready for the next indexing, and if a habit is made of always moving the sector immediately after indexing mistakes will not occur. Sometimes, when completing the indexing movement, the hole into which the plunger must be fitted is overshot and a small return movement is necessary. When this occurs, do not just return to the hole, but bring the crank right back about half a turn and go forward again. If this is not done any backlash in the worm and wheel may result in an error.

**Holding the work**

Work may be held on the dividing head in the following ways:
    (a) Between centres.

(*b*) One end in the chuck and the other supported by the tailstock centre.

(*c*) In the chuck.

(*d*) On an arbor fitted into the taper hole.

(*e*) By some special method adapted to suit a particular job.

The method of holding between centres adapts itself to shafts centred at each end, and to cylindrical work when it is held on a mandril. A better support for the head end is obtained when that end is held in the chuck, but care should be taken to ensure that the chuck jaws are true; otherwise the machined grooves will not be cut parallel with the axis of the work and their bottom faces will not be concentric. The same applies to work held entirely in the chuck. When there is a sufficient volume of work of the type which would normally be held on a mandril, with hole sizes coming within a few standard ranges, the expense of making a

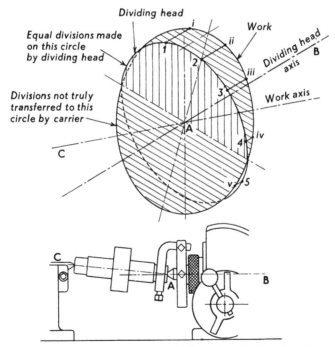

Fig. 150 Diagram to show fault in dividing when head and work are not co-axial (see text on p. 196)

special arbor is justified. Such an arbor fits into the taper hole in the spindle of the head and has its free end supported by the tailstock centre. The work is held by collars and a nut in the same way as the cutters on a milling machine arbor. This provides a support superior to a mandril because the taper end of the arbor is held rigidly in the head, and the work, being clamped by the collars hard up to a large shoulder, is held on its sides. Arbors of this type are useful for dealing with frequent batches of gears or milling cutters. A useful length for the parallel portion is about 200 mm to 250 mm with the end screwed for a fine threaded nut. If the hole in the work has to be ground after hardening, the arbor will have to be made to suit (0·15 mm undersize). The arbor will, of course, accommodate a string of components to the capacity of its length.

Work held normally between centres is driven by a carrier, somewhat similar to centre lathe practice except that the carrier shank must be held rigidly from moving in either direction. This is done by clamping the end of the shank between two jaws by a set-screw. Sometimes the carrier shank is provided with a ball which is clamped between half-round seatings incorporated in an arm which projects from the shank of the head centre. For accuracy in indexing with the work between centres it is important that the axis of the head should coincide with that of the work carried between the centres. The error involved by lack of alignment is shown exaggerated in Fig. 150. The jaws which actuate the end of the driving carrier are being indexed through equal divisions 1–2, 2–3, 3–4, etc., on circle 1, 2, 3, 4, etc., but these are being transferred to circle i, ii, iii, iv, etc., whose plane is not parallel, and it will be seen that the division i-ii (carrier vertical) will not be the same as iv-v (carrier horizontal). For accurate work, therefore, the

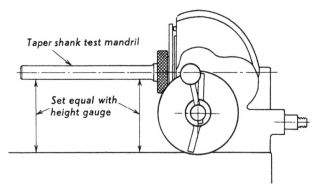

Fig. 151 Setting head horizontal

head axis must be set coincident with the line joining the centres, and the method of doing this depends upon whether the setting is horizontal, or at an angle (e.g. as it would be for milling the flutes of a taper reamer). When setting horizontal, first set the head itself by checking from the machine table to a taper shank test mandril held in the taper (see Fig. 151), then with a test bar between the centres adjust the tailstock until this bar also is parallel with the table. The quickest way of setting for an angular position is to set the head to the angle required and then place the tailstock along the table at the position it will be when the work is clamped between the two centres. Measure the distance between the two centre *points,* multiply by the sine of the angle and set the tailstock this amount above the head-stock (see Fig. 152).

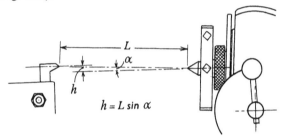

Fig. 152  Setting tailstock for milling taper work

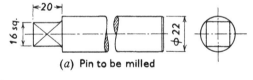

(*a*) Pin to be milled

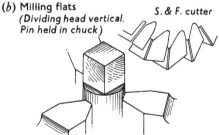

(*b*) Milling flats
(Dividing head vertical.
Pin held in chuck)

S. & F. cutter

Fig. 153  Milling square end on pin

## Examples of dividing head work

*Example* 3. *To mill the square end on the pin shown at Fig. 153(a).*

1. Put on the chuck and swing the head axis vertical. Fit a suitable side and face mill to the arbor. (The job could also be milled with an end mill held in the spindle.)

2. Clamp the job in the chuck leaving about 22 mm protruding and adjust the knee of the machine until the 20 mm length of square will be obtained.

3. Put on about $1\frac{1}{2}$ mm of cut with the cross-slide and traverse it across. Measure between the flat and the opposite circumference and continue the operation until this dimension is down to 19 mm.

4. Index 10 turns of crank, take off about 1 mm of the cut and then repeat the last operation.

(If the chuck is known to be true, once the first flat is correct, the same cut can be taken across for each of the four 90° positions. An eccentric chuck will necessitate a separate adjustment of the cut each time, but it will not affect the accuracy of the indexing.)

5. Index 10 turns and mill the third flat making it 16 mm from the first one.

6. Index and mill the fourth flat (Fig. 153(b)).

*Example* 4. *To mill the serrations on the shaft shown at Fig. 154(a).*

Since the *grooves* must be milled, their angle must first be determined, and by reference to Fig. 154(b):

$$\text{Angle AOB} = \frac{360}{48} = 7\frac{1}{2}°$$
and since angle ODC $= 37\frac{1}{2}°$ ($\frac{1}{2}$ angle of tooth)
$$\text{angle DCB} = 37\frac{1}{2} + 7\frac{1}{2} = 45°$$

So that the angle of the groove $= 90°$

If a double angle cutter of 90° (equal 45° on each side) is available with a 0·5 mm radius on its corners it can be set on the centre and the grooves milled as shown by (c).

It is unlikely, however, unless work of this type is a regular feature, that such a cutter would be available and an alternative method must be used. A suitable method would be to mill off-set, and use a small side and face cutter, but we shall require first to obtain a setting dimension.

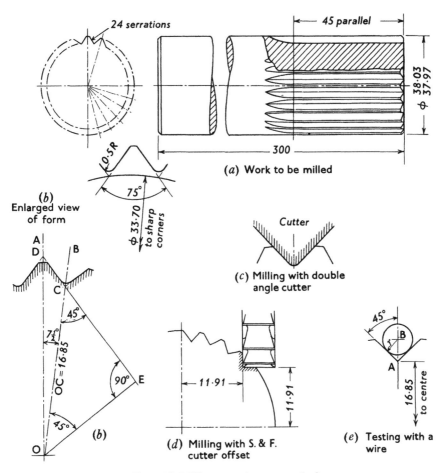

Fig. 154 Milling serrations on a shaft

In Fig. 154(b), if DC is produced to E, and OE drawn perpendicular,

$$\text{Angle OCE} = 45° \text{ and } OC = \frac{33\cdot70}{2} = 16\cdot85 \text{ mm}$$

Hence
$$CE = OE = OC \sin 45°$$
$$= 16\cdot85 \times 0\cdot7071$$
$$= 11\cdot91 \text{ mm}$$

The work should now be set between the centres of the dividing head, or one end held in the chuck and the other supported by the tailstock.

A small side and face cutter should be nicely sharpened up and a 0·5 mm radius stoned on the corners to be used for cutting. The cutter must now be set 11·91 mm from the centre of the work both in the cross and vertical directions as shown in Fig. 154(d). The cross setting may be done with a height gauge from the machine column face and the vertical setting from the machine table to a tooth of the cutter previously rotated to the extreme lowest position. Before these settings are made the cutter should be checked for true running on the top and sides of its teeth and if any discrepancy exists the settings should be made from the high localities.

When the settings have been made to satisfaction the cross and vertical slides of the machine should be locked, the dividing head set for 24 divisions and the table trip adjusted to trip the feed when the 45 mm parallel length of groove has been cut. The work can then be milled.

As a precaution, on the grounds of accuracy, a good method of working would be to off-set the cutter a few hundredths in excess of the finished position and after milling the first groove make a check on it with a wire measuring from the opposite extremity of the shaft. From Fig. 154(e), the distance from the centre of the shaft to a 3 mm wire placed in the groove would be:

$$16·85 + AB + r$$
$$= 16·85 + r \text{ cosec } 45° + r$$
$$= 16·85 + 1·5 \times 1·414 + 1·5$$
$$= 20·47 \text{ mm}$$

If the bar is 38 mm diameter the distance from the opposite diameter will be:

$$19 + 20·47 = 39·47 \text{ mm}$$

which may be checked with a micrometer. When the check has been made the cross and vertical slides may be advanced to remove any excess metal necessary to bring the groove to size.

*Example 5. To mill the teeth on the side and face cutter shown at Fig. 155(a).*

### 1st setting—Peripheral teeth

1. Press the cutter blank on a mandril and set between the centres of the head and tailstock.

2. Put a single angle cutter on the arbor. A cutter of about 75° or 80° should suit this job. (See *Senior Workshop Calculations*, p. 115 for determination of cutter angle.)

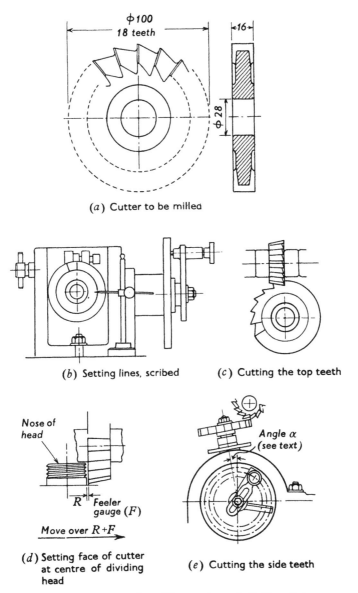

(a) Cutter to be milled

(b) Setting lines, scribed

(c) Cutting the top teeth

Nose of head

R Feeler gauge (F)

Move over R+F

(d) Setting face of cutter at centre of dividing head

Angle α (see text)

(e) Cutting the side teeth

Fig. 155 Settings for milling teeth on S. & F. cutter

Set the sector arms for suitable indexing:

e.g. $\dfrac{40}{18} = 2\dfrac{4}{18} = \begin{array}{l} \text{2 turns} + \text{ 4 holes in an 18 circle} \\ \quad\text{or} + 12 \quad\text{,,} \quad\text{,, a 54 ,,} \end{array}$

3. Since no rake is called for on the tooth front it must be cut on a radial line. Set the height gauge or scribing block to the centre height and scribe a horizontal line on one side of the blank. Index 1 tooth ($2\frac{4}{18}$ turns) and scribe another line. Index 10 turns to bring the second line to the top (Fig. 155(b)).

4. Set the cutter to the vertical line and put on a reasonable cut, but not enough to reach the first line. Traverse the cut through then index to the next tooth and take that through.

5. Now raise the table and deepen the cut until there is not more than a bare $\frac{1}{2}$ mm land at the top of the tooth.

6. When you are satisfied that the land is correct lock the knee and cross-slide, and cut the teeth all round. Use plenty of coolant on the cutter (Fig. 155(c)). (If you are doubtful of the indexing, go round it and mark each tooth position with the scribing block before commencing to cut.)

## 2nd setting—side teeth

For holding the blank in this operation a stub, taper shank arbor will be necessary, the blank locating on a short end fitting in its bore and being held by a set-screw and washer.

1. Fit the blank to the arbor and clamp it up tightly. Swivel the body of the dividing head to the correct angle (see next paragraph). Place an angle cutter on the machine arbor and clamp it up. (A cutter of 80° or 85° should be suitable.)

2. Set the vertical cutting edges of the angle cutter on the centre line of the dividing head. One method of doing this is to set with feeler gauges from a diameter (e.g. the nose) of the head spindle then move the table over by half the diameter used plus the thickness of the feeler (Fig. 155(d)). Lock the cross-slide and fit the job and its arbor to the spindle of the head.

3. Start up the machine and wind the table so that the work is close to the cutter. Carefully index round until the cutter just scrapes the face of one of the teeth previously cut. Put on a small cut and traverse it through.

4. Index to the next tooth, take the cut through and then raise the table until the land is the same as that on the top teeth.

5. Lock the knee and cut the teeth all round the blank (Fig. 155(e)).

**3rd setting**

Remove the blank from the arbor and reverse it for cutting the teeth on the other side.

Cut these teeth by repeating operations 3 to 5 above.

**Setting for cutting side teeth**

If, when milling the side teeth, the face of the blank is set horizontal, the lands of the teeth would not come parallel owing to the reduction in diameter which occurs as the cutter travels towards the centre. To compensate for this the blank must be tilted to an angle $\alpha$ (Fig. 155(e)), so that the flute is deeper at the outside diameter than at points nearer the centre. The conditions necessary for obtaining a parallel tooth land are shown at Fig. 156, and all lines on the milled profile must converge to the centre point O. In the diagram: $\theta =$ the angle subtended by a tooth at the centre, i.e.

$$\theta = \frac{360°}{\text{No. of teeth in cutter}};$$

$\phi$ is the angle of the fluting cutter and $\alpha$ is the tilt of the blank (i.e. inclination of the dividing head from the vertical).

Let AC be drawn perpendicular to OD, and CB perpendicular to OE. Then triangle ABC is right-angled at C, and angle B $= \phi$. (In this and the following diagrams 90° angles are indicated thus $\llcorner$ .)

$$OA = \text{radius of cutter blank } (r).$$

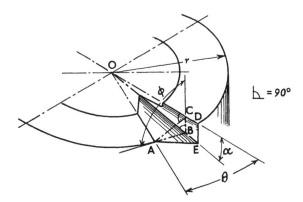

Fig. 156  Milling side teeth in cutter

Then
$$\sin \alpha = \frac{BC}{OC} = \frac{AC \cot \phi}{OA \cos \theta} = \frac{OA \sin \theta \cot \phi}{OA \cos \theta}$$
$$[\text{since } AC = OA \sin \theta]$$
$$= \frac{\sin \theta \cot \phi}{\cos \theta} = \tan \theta \cot \phi \quad \left[ \text{since } \frac{\sin \theta}{\cos \theta} = \tan \theta \right]$$

Hence   **$\sin \alpha = \tan \theta \cot \phi$**
For the cutter we have been considering:

$$\theta = \frac{360}{18} = 20°; \ \phi = 80°$$

Hence
$$\sin \alpha = \tan 20° \cot 80°$$
$$= 0·363\ 97 \times 0·1763 = 0·064\ 168$$
from which
$$\alpha = 3° 40'$$
and the dividing head would have to be set at
$$90° - 3° 40' = 86° 20'.$$

### Fluting angular cutters and taper reamers

A similar set of conditions to those we have just discussed governs the fluting of tapered cutters. If the flute is cut parallel with the conical surface of the blank, the lands at the top of the teeth will not be parallel, too much metal being removed from the smaller diameter of the taper. The correct method of working in such cases is to set the blank at such an

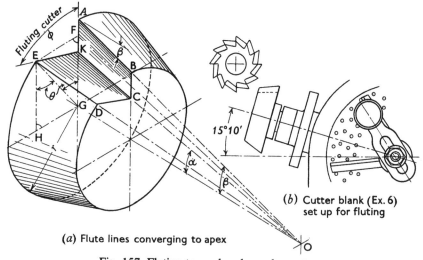

(a) Flute lines converging to apex

(b) Cutter blank (Ex. 6) set up for fluting

Fig. 157 Fluting tapered and angular cutters

angle that all lines on the flute converge to the apex of the taper. This is shown at Fig. 157(a), where O is the apex and ABCDEK the flute, the lines AB, KC and ED all meeting at O, when produced. G is the centre of the large diameter of the blank, EH is parallel to AG and EF is perpendicular to AG. Then the half taper angle $\beta$ is the angle AOG and the angle subtended by one flute at the centre ($\theta$) is the angle AGE = GEH. GE = rad of cutter ($r$). The angle to which the cutter must be set is angle KOG, and a simple approximate solution for calculating it may be obtained if we assume that angle AKE is the angle $\phi$ of the fluting cutter. Actually, this is not quite true, as its real angle is that of the flute measured perpendicular to KC, but the error will not be great for reasonably slow tapers.

If $\alpha$ is the setting angle:

$$\tan \alpha = \frac{KG}{GO} = \frac{KG}{AG \cot \beta} = \frac{KG}{r \cot \beta}$$

$$= \frac{FG - FK}{r \cot \beta} = \frac{EG \cos \theta - FK}{r \cot \beta} = \frac{r \cos \theta - FK}{r \cot \beta}$$

$$= \frac{r \cos \theta - EF \cot \phi}{r \cot \beta} = \frac{r \cos \theta - HG \cot \phi}{r \cot \beta}$$

But
$$HG = EG \sin \theta = r \sin \theta$$

$$= \frac{r \cos \theta - r \sin \theta \cot \phi}{r \cot \beta}.$$

The $r$'s cancel out, which leaves us with

$$\mathbf{tan}\ \alpha = \frac{\cos \theta - \sin \theta \cot \phi}{\cot \beta}$$

*Example* 6. *The setting angle for milling a* 70° *blank with* 18 *teeth. using a* 60° *fluting cutter.*

Here
$$\theta = \frac{360}{18} = 20°; \phi = 60° \quad \text{and} \quad \beta = 90° - 70° = 20°$$

$$\tan \alpha = \frac{\cos 20° - \sin 20° \cot 60°}{\cot 20°}$$

$$= \frac{0 \cdot 9397 - 0 \cdot 342 \times 0 \cdot 577}{2 \cdot 747}$$

$$= 0 \cdot 271$$

From which $\alpha = 15° \ 10'$.

The setting is shown at Fig. 157(b).

**Cutting clutch teeth.** The face teeth of clutches may be cut on the milling or shaping machine, using the dividing head with its spindle set vertical. Dog clutches usually have teeth with square faces, and in cutting these a side and face mill may be used with one face set on the centre. One face of each tooth is then milled and then the cutter must be set over until its other face is central, after which the uncut face of a tooth is indexed into position and the blank indexed round for finishing the other side of each tooth. This is shown at Fig. 158(a). If the clutch has an odd number of teeth, and there is sufficient clearance not to cut off the edge of the next tooth, each cut may be taken straight through, finishing the faces of two teeth at one passage (Fig. 158(b)). If the diameter of the clutch is small and the teeth fairly deep, the cutter may run into the opposite tooth before it has made sufficient travel to complete its cut. The remedy for such cases is to take the job on to a shaping machine or use an end mill on a vertical miller.

### Saw tooth clutches

The setting for milling or shaping the teeth of saw tooth clutches is much the same as we have sketched for cutting the side teeth of side and face mills (Fig. 156). For the best conditions of engagement these teeth should be cut so that the inclinations of the tops and bottoms of the teeth are equal, as shown at Fig. 159(a). With this arrangement two sets of mating

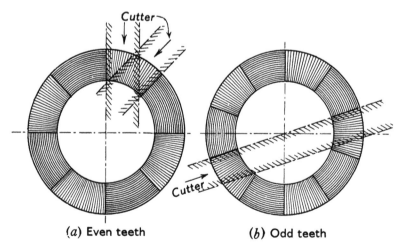

(a) Even teeth          (b) Odd teeth

Fig. 158 Milling dog clutch teeth

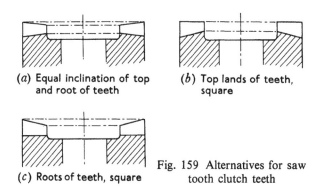

(a) Equal inclination of top
and root of teeth

(b) Top lands of teeth,
square

(c) Roots of teeth, square

Fig. 159 Alternatives for saw
tooth clutch teeth

teeth will engage correctly over their whole surface. If the teeth are cut as at (b) to give square, parallel lands to the top of the teeth or as (c), where the cutter is fed perpendicular with the axis, the two clutch members would not mesh completely because certain parts of the teeth would strike bottom before the whole tooth was in engagement.

The considerations necessary for cutting the tooth form shown at Fig. 159(a) are shown at Fig. 160(a).

KH is the axis of the clutch and ABCDEF the boundaries of a tooth, AB being parallel to KH. Point G is the mid-point of AB, GO being projected across to the axis to give the point O, midway between K and H.

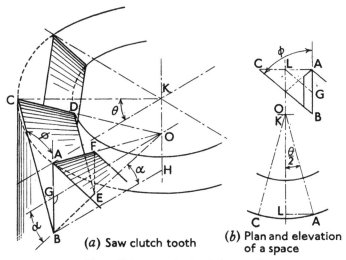

(a) Saw clutch tooth

(b) Plan and elevation
of a space

Fig. 160 Saw tooth clutch (see text)

Then the conditions are such that every line on the tooth must converge to O.

In the diagram, $\phi$ is the angle of the tooth, $\theta$ the spacing angle $\left( \dfrac{360°}{\text{No. of teeth}} \right)$, and $\alpha$ the angle to which the blank must be set for cutting the teeth.

$$\tan \alpha = \frac{AG}{GO}$$

and from (b):   $AL = AK \sin \dfrac{\theta}{2}$   and   $AG = AL \cot \phi$

Hence   $AG = AK \sin \dfrac{\theta}{2} \cot \phi$

so that   $$\tan \alpha = \frac{AK \sin \dfrac{\theta}{2} \cot \phi}{GO} = \sin \frac{\theta}{2} \cot \phi$$

since $AK = GO$.

*Example 7. The setting angle $\alpha$ for cutting 8 teeth in a saw tooth clutch with a tooth angle of 60°.*

Here   $$\theta = \frac{360}{8} = 45° \quad \text{and} \quad \phi = 60°$$

$$\tan \alpha = \sin 22\tfrac{1}{2} \cot 60 = 0 \cdot 3827 \times 0 \cdot 5773 = 0 \cdot 2209$$

From which   $\alpha = 12° \, 28'$

i.e. the dividing head setting $= 90° - 12° \, 28' = 77\tfrac{1}{2}°$ approx.

## Gear cutting*

Gears may be cut on the dividing head by using a special form cutter which gashes out the tooth spaces. This method is not used when large quantity production of gears is concerned, but for odd jobs it is extremely valuable and is capable of fulfilling a wide range of needs, provided a good selection of gear cutters is available. In order to cut a full range of gears under this system eight cutters are required for each pitch. This is a compromise, as the profile of a gear tooth is different for every variation in diameter of the wheel, but the set of eight cutters to which we refer strikes an average

* Gears and their calculations are discussed in the author's books on *Workshop Calculations.*

and gears cut with them function satisfactorily. The set of eight cutters necessary to deal with each pitch on the involute system is as follows:

No. 1 will cut wheels from 135 teeth to a rack.
,, 2 ,, ,, ,, ,, 55 ,, ,, 134 teeth
,, 3 ,, ,, ,, ,, 35 ,, ,, 54 ,,
,, 4 ,, ,, ,, ,, 26 ,, ,, 34 ,,
,, 5 ,, ,, ,, ,, 21 ,, ,, 25 ,,
,, 6 ,, ,, ,, ,, 17 ,, ,, 20 ,,
,, 7 ,, ,, ,, ,, 14 ,, ,, 16 ,,
,, 8 ,, ,, ,, ,, 12 and 13 ,,

One of these cutters is shown at Fig. 120, and its method of application at Fig. 161. The gears in most general common use have module ($m$) pitches ranging from about 1 mm $m$ (tooth 2 mm high) to 16 mm $m$ (tooth 32 mm high). For teeth larger than about 12 mm $m$ the tooth dimensions are often expressed in terms of their circular pitch ($p = \pi m$ mm).

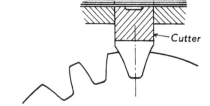

Fig. 161 Cutting a gear with a form cutter

When cutting gears on the dividing head the important points to observe are: (1) to ensure that the teeth are cut concentric with the bore of the gear, (2) to sink the cutter in the correct depth, and (3) to set the cutter accurately on the centre of the gear. The precautions in the case of 1 are to ensure that the mandril upon which the gear is mounted runs true or, if the work is held some other way, to set it up with a clock indicator from the bore if possible. The depth to which a cutter must be sunk is generally stamped on it but, if not, it can be obtained by multiplying 2·25 by the module. Thus for a $6m$ cutter the whole depth would be $2·25 \times 6 = 13·50$ mm. It is usually advisable not to cut to the full depth at once, but to go round leaving a small amount to be taken out by a second finishing cut. Gear cutting provides an example where the value of the graduations on the sleeve of the knee elevating handle should be known with confidence. The depth of tooth is important and is not easy to measure if one cannot rely on the vertical feed indicated when sinking in. We have already discussed one or two methods of setting the cutter on the centre of the work. There are many more, and we can leave the reader to discover them for himself, but in passing we might remind him that if the cutter is rotated and passed from side to side across the top of a circular job, the knee meanwhile being gradually raised, the cutter teeth will eventually scratch the surface of the work. This scratch will be exactly above the centre.

*J. Parkinson & Co, Ltd*

Fig. 162 Rack cutting with universal attachment and special vise

The general method to be adopted for cutting a gear is the same as we have discussed in connection with the peripheral teeth on the cutter in Example 5 above. If the gear is too large to be accommodated by the head centre height the head and tailstock may be packed up, or the head may be rotated to bring its axis vertical. The gear is then held with its plane horizontal and the cut traversed by elevating the knee, with the table and cross-slide locked. Jacks will probably be necessary for support against the thrust of the cut.

**The universal and rack cutting attachment**

For cutting teeth or grooves perpendicular to the length of long bars, it is necessary for the milling cutter to rotate on an axis parallel with the table, as by working in the normal manner it would not be possible to accommodate the work, or obtain sufficient cross-slide movement to cover it. The rack cutting attachment is a head which is bolted to the front of the machine and is provided with a spindle and short cutter arbor rotating on an axis parallel with the table. The drive is taken from the machine spindle, and gearing inside the attachment converts the rotation to the direction of the head spindle axis. When dealing with jobs of this kind the work is secured with its length parallel with the table and the cut is traversed by moving the cross-slide. Lengthwise spacing of teeth or grooves is carried out by making use of the graduated sleeve of the table lead-screw or, if this has no such fitment, a method as suggested in Fig. 169 may be used. The table should be locked whilst the cut is taking place. A diagram of the *universal milling attachment* cutting a rack is shown at Fig. 162. This attachment is more versatile than the rack cutting attachment since its horizontal spindle may be rotated to any angle. This feature facilitates its use for spiral milling where the helix angle is beyond the capacity given by the rather restricted swing of the machine table. As well as having a horizontal spindle the head will take cutters for vertical milling so that it may be used in lieu of the vertical attachment shown at Fig. 164.

### Vertical Milling

When the reader has had experience of working with an end or face mill on a horizontal machine he will think how much more convenient it would be if the cutter were above instead of at the side of the work. This is the arrangement on the vertical machine which has knee, cross-slide and table similar to the horizontal miller, but has a spindle disposed vertically above the table.

The applications and technique of this machine differ considerably from the horizontal miller. It uses no supported arbor and cutters such as we have just been discussing, but does most of its work with end and face cutters. Occasionally, but not often, a side and face mill mounted on a short stub arbor is found useful for cutting grooves or facing underneath. For much of the work it produces the machine is closer akin to the shaper, planer, slotter family than to its namesake, but it is nevertheless still a milling machine, and as such is able to deal with a range of work which could be done on the other four machines but which it handles more efficiently. In addition, the vertical miller can undertake jobs which are essentially its own and which could not be machined economically by any other method.

### The vertical machine

The chief features of a vertical milling machine are shown at Fig. 163, the machine shown having a table with a working surface of 1000 mm × 315 mm, 12 speeds and a 3 kW reversing driving motor. The spindle head is on its own slides and has 70 mm of movement actuated by the handwheel shown. Automatic feed (18 feeds, 16 mm to 800 mm/min) is provided on the table and cross-slide, the movement of all slides being indicated by graduated sleeves.

*Rockwell Machine Tool Co, Ltd*

Fig. 163 Fritz Werner VF 1·5 vertical milling machine

### The vertical head

Vertical milling may be carried out on a horizontal machine by making use of a vertical head which bolts against the front face of the column, being supported by the overarm, and has its spindle driven through right-angle gearing by some form of adaptor which fits the end of the machine spindle. Naturally, the scope for vertical milling of a horizontal machine with vertical head is more restricted than if a good vertical machine is available, but not all shops have sufficient vertical work to warrant the expense of a machine, and for their purpose the head is sufficient. As compared with the vertical machine, the head is less rigid and its spindle much less

generous in its dimensions. The head has no independent vertical movement for its spindle as the vertical machine, and cut has to be put on by
raising the knee. It has one useful feature, however, which is not shared
by many vertical machines, this is, that it may be swung round so that the
spindle axis is inclined at angles other than the vertical, and for some jobs
this is a valuable feature. An illustration of a vertical head on a horizontal
machine is shown at Fig. 164.

*J. Parkinson & Son, Ltd*

Fig. 164  Vertical milling head

**Equipment for vertical milling**

For the vertical machine to be used to its maximum advantage it should
be supplied with a good range of end and face mills. End mills should range
from 6 mm diameter to about 40 mm in the solid variety, with a selection
of shell end mills for diameters up to about 100 mm. The smallest face mill
might be about 150 mm diameter, and these should go up in 50 mm steps
to 300 mm or 350 mm, depending on the size of the machine. A selection
of two-lipped slot drills will be useful for getting out keyways and slots
and, in addition, stub, angular and tee slot cutters will be needed occasionally. For holding a side and face mill for various purposes a stub arbor
should be available. Drills will be needed for some work, but these will, of
course, be available from other sources.

The equipment of the machine often includes a circular table such as we discuss later with relation to the slotter, but as a self-contained unit which may be bolted to the table. Some machines have provision for automatic feed for this, but more often than not it must be hand fed. In its application, it fulfils much the same need as the one used with the slotter, enabling circular profiles to be machined by cutting on the side of end mills. Other accessories for use on the machine will be a selection of vises, a dividing head occasionally and a good supply of bolts and tackle for setting up, similar to those employed on planing and boring machines.

**Examples of vertical milling**

*Example 8. Operations for milling the bracket and cap shown at Fig. 165(a).*

The fitting of the studs and boring the main hole will be carried out after milling has been completed.

**Cap**

1. Check up for machining allowance, clamp in the vise, setting the face parallel with the table and the long edges of the base parallel with the table tee slots.
2. Using a shell end mill about 50 mm to 75 mm diameter, rough mill each surface and each side of the tenon which fits into the base. Keep the tenon faces central with the bore. When about the last $\frac{1}{2}$ mm remains on each lower surface put on the cut and lock the knee and spindle slide so that the surfaces on each side will be at the same level. Finish the surfaces and width of the tenon sides (Fig. 165(b)).
Finish the two narrow upper surfaces.
3. Hold in vise, making sure that machined face is flat against fixed jaw and set one of the tenon sides square with the table. Rough and finish the side of the boss (Fig. 165(c)).
4. Reverse in vise, set the previously machined face level on base of vise jaws and machine opposite face to width.

**Base**

1. Measure up the casting to check if there is sufficient machining allowance. Hold in the vise and jack under each end. Set the rough base parallel with the machine table, using a scribing block.
2. With a 150 mm or 200 mm face mill, put on sufficient cut to get below the scale and traverse it across. Add additional roughing cuts as

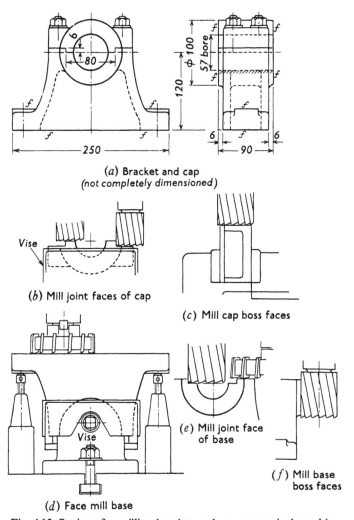

(a) Bracket and cap
(not completely dimensioned)

(b) Mill joint faces of cap

(c) Mill cap boss faces

(d) Face mill base

(e) Mill joint face of base

(f) Mill base boss faces

Fig. 165 Settings for milling bracket and cap on vertical machine

necessary, leaving a small amount for a final finishing cut. For the finishing cut use a fine feed (d).

3. Remove the vise and clamp the job on its base. Set the side edge of the base parallel with the table tee slots. Using the face mill rough and finish the top.

4. Remove the face mill and put in the shell end mill. Rough the step

on each side, keeping the cored hole central with the tenon faces and leaving about ½ mm on each face for finishing. Finish the steps to size and try in the cap until it is a nice push fit (e).

5. Cutting with the side of the end mill rough and finish the boss end faces, making the width the same as the cap (f).

6. Face up the end bolt bosses with the same cutter.

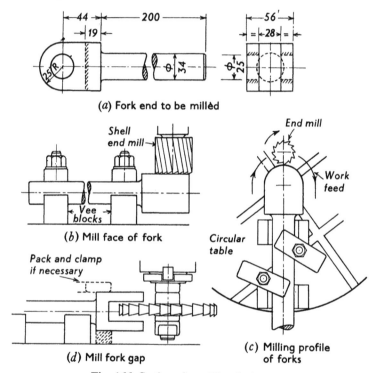

(a) Fork end to be milled

(b) Mill face of fork

(c) Milling profile of forks

(d) Mill fork gap

Fig. 166 Settings for milling fork end

*Example 9. Machining operations on the jaws of the component shown at Fig. 166(a).*

The component is to be made from a forging and when coming for milling will have the 34 mm diameter shank turned. The 25 mm hole will be put in after milling the jaws.

1. Support and clamp the component on vee blocks and skim up one of the outside faces of the jaws, using a shell end mill (b). Mark off the profile on the machined face.

2. Fit the circular table, and again clamping on vee blocks set the centre of the marked-off radius concentric with the table. The previously machined surface should be set horizontal.

3. With the shell end mill, rough the profile to within about $\frac{1}{2}$ mm of the marking out. Use table traverse for straight portion and circular table traverse for the radius (c).

4. Fit a side and face cutter on a stub arbor and get out the gap in the jaws. To machine this central with the shank first machine one face, measuring it with a height gauge from the table. It should be $3\frac{1}{2}$ mm towards the centre of the pin from the outside diameter of the shank. Finish the gap to a plug or block gauge (d).

5. Replace the shell end mill and finish the profile.

6. Mill the top surface to size (b).

7. Turn round in the vee blocks and set machined face parallel with table. Rough and finish the opposite face.

### Miscellaneous set-ups

A few illustrations showing jobs being machined, mostly by vertical methods, are shown at Fig. 167. The operation shown at (a) and (b) is interesting, as it shows vertical and horizontal methods applied on the same machine without disturbing the job. At (a) the base of the casting is machined with the face mill and then the vertical head is swung into position so that the vertical slot may be end-milled as at (b). After our discussions regarding the value of carrying out the maximum of work at one setting, it will be seen that adopting this method ensures squareness of the vertical slot with the base. At (c) is shown the slot-drilling of two keyways in a sleeve. The 25 mm diameter slot drill operates at 400 rev/min and a feed of 38 mm per minute. An interesting point is the gauge shown for setting the first machined keyway when rotating the work for cutting the second slot. The operation at (d) is that of milling the angular faces on the casting of a vertical milling attachment to provide the surfaces where it fits the overarm. The gap is cut from the solid, the preliminary operation being to mill a slot, followed by the operation shown. Milling the webs of a crankshaft is shown at (e). The 200 mm face mill takes a roughing cut 12 mm deep at a feed of 50 mm per minute, and a 3 mm finishing cut at 100 mm per minute feed. The crankshaft is supported in a fixture which provides supporting bearings at the two ends and in the centre. One end and the centre bearing can be seen in the diagram.

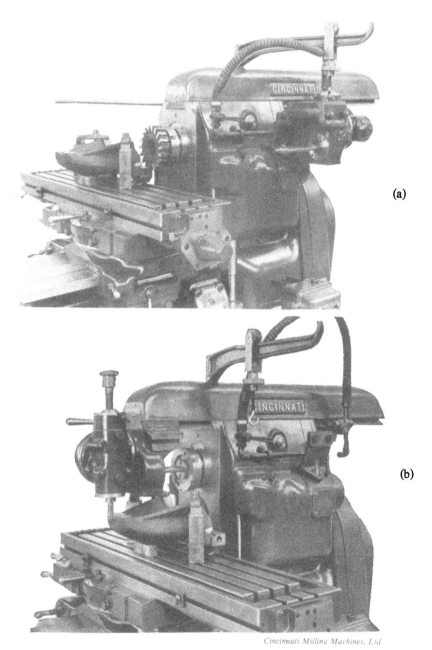

(a)

(b)

*Cincinnati Milling Machines, Ltd*

Fig. 167 Vertical milling operations (see text)

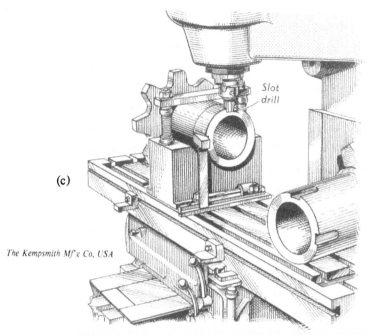

(c)

Slot
drill

The Kempsmith Mf'g Co, USA

(d)

James Archdale & Co, Ltd

Fig. 167 (*continued*).

(e)                    *Kendall & Gent* (1920), *Ltd*

Fig. 167 (*continued*).

### Boring on the milling machine

The milling machine, and particularly the vertical type, lends itself to certain classes of boring work. When no other machine of a suitable type is available many jobs may be carried out by a suitable set-up on the vertical machine and, to a lesser extent, the horizontal machine with vertical or universal attachment.

A particular class of boring for which a good vertical miller is suited is called *jig boring*. This class of work occurs commonly in connection with the construction of jigs and fixtures, and consists of boring holes in castings, plates, etc., which form the components of equipment used in the factory to set and hold the work during the processes of manufacture. The holes occurring in jig boring are not usually long ones such as the bores of long cylinders, etc., but are reasonably small and short, located in plates and in short bosses incorporated on the cast form of the fixture or in brackets attached thereto. Generally, the important requirements of such holes are their positional accuracy and their size, although often the size is subsidiary to correctness of location, as bushes which must be ground to fit the holes may be finished to suit their bore. It will be realised that when a jig is to be used in locating the work and guiding the cutting during the machining of other components, there can be no doubt as to whether the jig need be just correct or not; it must be *right* to within the smallest tolerance that human skill can uphold. We have already discussed the application of fixtures in connection with milling and will later consider drilling jigs. In addition to such purposes, equipment is often necessary for turning, gauging, assembly, boring, holding and locating whilst brazing or welding is carried out and for such other purposes as may be required to meet the need. In view of the extreme accuracy necessary and of other factors involved such as setting up, measuring, etc., jig boring constitutes one of the most skilled operations in the toolroom and is only entrusted to the most skilled and reliable men. We have already discussed button boring in the chapter on the lathe. This is a form of precision boring which comes in the category of jig boring, but as its application is limited to holes in flat plates, and in work which can conveniently be attached to the faceplate of a lathe, its range is limited. Brackets, castings and bulky work cannot be swung in this way and must be bored with a revolving tool whilst fixed in a stationary position

**Tool head**

When boring is done on the vertical miller the machine should preferably be of a type on which the spindle has its own vertical slide with power-fed movement. This not only facilitates the feeding of the boring tool, but the lighter spindle slide is a more satisfactory and sensitive method of feed-

*George Richards & Co, Ltd*

Fig. 168 Adjustable high-speed boring head

ing the rather delicate boring tool, and gives better working conditions than is possible by feeding the heavy and unresponsive knee and table unit. To carry the single-point boring tool and provide for its radial adjustment to vary the size of hole bored a special head is necessary. A diagram of one type of such attachment is shown at Fig. 168.

### Work setting and hole location

As a general rule, when work comes for boring, all the necessary flat machining and turning will have been done, so that there will be suitable finished surfaces for location. The drawing will specify from which surfaces the relative location of the holes is important, and it will also give some of the centre dimensions for the holes. If the dimensions given on the drawing are not helpful for setting or checking, then the dimensions required will have to be calculated from those given. It can be assumed that the drawing dimensions completely tie down the position of a hole even although they may not be the most suitable ones from which to locate and bore it. Care is necessary to avoid using unimportant surfaces as datum, or as intermediaries for the purpose of obtaining a relationship. In the same way errors will be avoided by keeping to one or two surfaces only for the purpose of reference, e.g. on a bracket with a machined base and two machined ends, work from the base and one end only for the location of the bored holes in the bracket. The drawing will not specify definite instructions about holes being parallel with certain faces or with each other, being square with certain faces, being round, parallel, and so on; but we have said enough on these points in former chapters to render further comment unnecessary.

If the alignments of the vertical machine are correct it will bore holes perpendicular with the table surface and parallel with a surface which is set square with the table. If the cross-slide is kept locked and two holes bored by locating their centres by a movement of the table only, their centres will lie on a line parallel with any surface which was originally set parallel with the table tee slots. If, after the second of the above holes has been bored, the *table* is kept locked in the same position and the *cross-slide* moved, a third hole may be bored on a centre line at right-angles to that joining the centres of the first two, and so on. Considerations such as these which apply to milling, just as much as to boring, must guide the reader in setting and arranging the progress of his work and, as we have previously stressed, it is by such precautionary forethought and planning that the skilled man's work seems to go through so easily.

The centre locations of bored holes must be fixed by the two perpendicular movements of table and cross-slide, so that their centre distances, if not set out on the drawing in this fashion, will require conversion, so as to read from two axes at right-angles (rectangular coordinates). After one hole has been located and bored the preliminary position of the next may be obtained by a careful movement of the slides necessary by using the graduated sleeve or dial to control the distance moved. This should not be relied upon as accurate, but a final check made with test bars after a preliminary undersize boring of the second hole. Sometimes it may not be possible to carry out a check during boring and the initial movement must be relied upon as final. Such a state of affairs renders the accuracy of the movement important, and if the screws and dials are unreliable or the dials non-existent, other means must be adopted for controlling the movement. One such method, which may suggest others to the reader, is shown at Fig. 169.

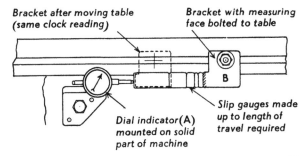

Fig. 169  Method of obtaining an accurate movement to a machine table

To a suitable fixed point on the stationary machine casting is fixed a dial indicator A. Attached to the moving slide at a convenient point with relation to the pointer of the indicator is a measuring stop B. The space between the measuring face of B and the plunger of the dial gauge may be measured with slip gauges or by any convenient method, the dial gauge being taken to a zero each time a setting is made. It would serve of course, if a solid measuring point were substituted for the dial gauge, but the latter ensures uniformity of measuring pressure and can itself be used for indicating short movements, or small differences from the length set between the points. When boring, it should be remembered that a movement of the tool varies the size bored and has no effect on the centre position, whilst moving the work causes a change in the centre position at which boring is taking place. When holes have to be bored in castings it will generally be found that those whose finished diameter is 35 mm or over will be cored about 4 mm to 6 mm less in diameter than when finished. In other cases the metal will be solid and the hole must be drilled before a start can be made with the boring tool. When a hole is cored and the boring happens to be somewhat out of centre with the rough hole, some tool spring may take place on the side where the deepest cut occurs. This will mean that the first boring from the rough may be neither round, parallel nor concentric with the travel of the tool, and a second or third passage of the tool may be necessary before an accurate hole is obtained.

*Example* 10. *Boring operations for the bracket shown at Fig.* 170(a).

(The two large holes will probably be cored out, and the boss for the vertical hole solid metal.)

**1st setting**

1. Fasten an angle plate to the table of the machine and set its face perpendicular with the length of the table. Clamp the base of the bracket to the angle plate and set so that the level outer boss faces are uppermost, and parallel with the machine table. Support under the central projecting lug and clamp lightly on to the support. The tenon slot in the base should now be perpendicular with the table and this might be tested by checking with the height gauge from the face of the vertical knee slides or from some surface known to be perpendicular with the table. If the edge of the base on the same side as the 32 mm hole lug is machined it may be rested in contact with the table and so serve as both support and location. The set-up is shown at Fig. 198(b).

2. For setting the work into position for boring the first hole, a test bar with a taper shank fitting the machine spindle is useful. This might be about 150 mm long and 25 mm diameter on the parallel portion which projects from the spindle. Fit this or something similar in the spindle and set its centre 144·00 mm from the angle plate and 67·00 mm from the tenon slot centre, working with height gauge from angle plate and vertical face of machine respectively. (Check the tenon slot for width before taking readings from one of its sides.) Working crossways with the height gauge is faciliated by having a plate which can be bolted to the edge of the table for providing a vertical datum surface (see Fig. 170(c)).

3. Fit the boring head and take a roughing cut through the 48 mm hole, keeping below the finished size and opening out until a test plug can be

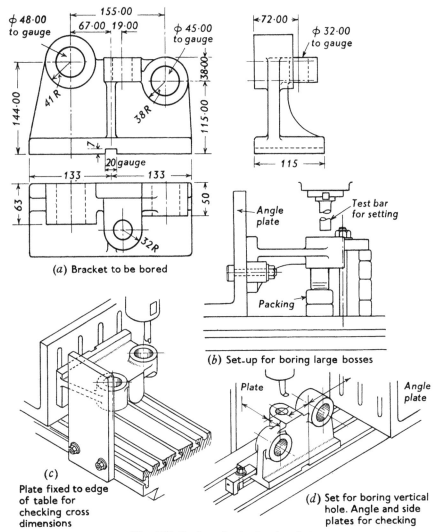

(a) Bracket to be bored

(b) Set-up for boring large bosses

(c)
Plate fixed to edge
of table for
checking cross
dimensions

(d) Set for boring vertical
hole. Angle and side
plates for checking

Fig. 170 Settings for boring bracket

fitted 46 mm or 47 mm. Fit the test plug and check the centre position. Make any positional adjustments to the machine slides necessary to correct the centre position and bore the hole out to size. (If still in doubt carry out a second check after the reboring following the first adjustment.)

4. Move the cross-slide through 155·00 mm and the table 29·00 mm

(144−115) using the graduated sleeves. (Accuracy will be promoted here if, when the two slides were finally brought into position for the first hole, they were moving in the same direction as they have just been moved to bring the second hole under the spindle. Working this way avoids trouble with the backlash in the screw.)

5. Rough bore the second hole and when a test plug can be inserted check the 155·00 mm and the 115·00 mm dimensions, working with the height gauge crosswise and longitudinally. If these are correct check the hole centre distance $[=\sqrt{155^2+29^2}]$ to make sure. Adjust to correct if necessary and finish bore to a 45 mm plug gauge.

### 2nd setting

1. Clamp the job to the table on its base and set the level boss faces parallel with the tee slots. Put the setting test bar in the spindle and set it 19·00 mm from the centre of the tenon slot and 72·00 mm from the boss faces. (The 19·00 setting can be measured from the face of an angle plate bolted across the table and the 72·00 mm from the vertical plate previously used.)

2. Drill the hole first about 12 mm and then 30 mm. Fit the boring head, take a trial cut through and check the hole position. Correct the setting if necessary and open out the hole to the 32 mm plug gauge (Fig. 170(d)).

# 7 Shaping, planing and slotting

These machines form a family, the main function of which is the production of flat surfaces. The plane of the machined surface is obtained by the combination of a line tool cut with a perpendicular feed of the tool or work, as discussed on page 103. The three types of machine serve an important purpose in a general machine shop—the shaper, particularly, on account of the quickness with which the work and tool can be set up. All the machines are versatile in respect of the range of work that may be handled and they can all be relied upon to produce flat surfaces of good accuracy. This achievement, moreover, is obtained at very low cost in tools, since the machines require only the simplest of single point tools.

As we have already discussed the general construction and fundamental uses of the crank shaper,* we will go on to consider the more advanced and specialised aspects of shaping.

### The hydraulic shaping machine

The application of a fluid under pressure to actuate a piston or plunger for the purpose of obtaining a linear motion has been developed in the design of various machine tool movements. The idea is far from being new, since hydraulic power, using water as the medium, has been applied to presses and other equipment since very early days of engineering. Compared with mechanical means of producing a motion, the hydraulic method, although tending to be more costly, offers under suitable conditions various advantages. The most important of these are (1) simplicity, (2) flexibility, (3) operation and control, (4) smoothness, (5) fewer working parts, and freedom from wear, (6) a greater measure of safety, since the introduction of relief valves allows the pressure to be diverted in the event of an obstruction in the movement. On the shaping machine the ram, with its longitudinal motion, is the principal element, so that the shaper is a deserving case for the adoption of any method of driving which might

* *Workshop Technology*, Part 1, Chap. 12.

*The Butler Machine Tool Co, Ltd*
Fig. 171 Hydraulic shaping machine, 460 mm stroke

possess advantages over mechanical actuation. A diagram of the 'Butler' 460 mm stroke hydraulic shaper is shown at Fig. 171. In external appearance and in its movements the machine has very similar characteristics to the crank-driven machine, but after that the similarity ends, for the inside of the body, instead of containing machinery, is mostly taken up with oil pipes and control valves.

The $5\frac{1}{2}$ kW motor, mounted at the back, drives two constant-delivery pumps, one for the ram and one for the feed motions. The speed of the ram is governed by a flow control valve which allows two ranges of speed. On the high range the forward speed can be varied steplessly from 0 to 40 metres per min and the return speed ratio is $1\frac{1}{8}$ to 1. The low range setting gives a forward speed range of 0 to 21 metres per min and a return speed ratio of $2\frac{1}{8}$ to 1. The change from one range to the other is effected by a lever which operates the main valve through a pilot valve. An additional finger-operated lever serves to start and stop the machine, and the stroking and reversal of the ram are controlled by adjustable dogs which trip another valve situated on the top of the machine. The feeds are obtained through a variable stroke hydraulic piston, which actuates a roller ratchet, the

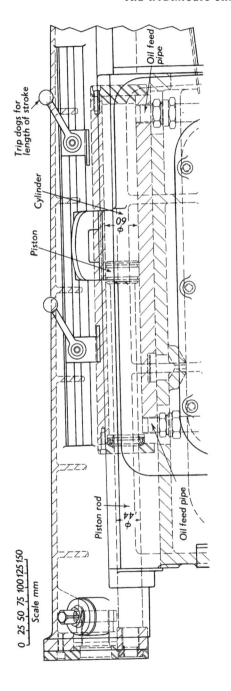

*The Butler Machine Tool Co, Ltd*

Fig. 172  Elevation of upper part of body and ram of hydraulic shaping machine

effective movement being controlled by a handwheel mounted on the saddle. This arrangement provides a feed, infinitely variable from zero to 2·5 mm, which can be applied to move the table horizontally or vertically. Most of the above features can be picked out on Figs 171 and 172, and diagrams of the hydraulic circuit are shown at Fig. 173. These show for the low range at (a) the cutting stroke and at (b) the return stroke, and, whilst we cannot go into details as to what takes place inside the control valves, the diagrams will convey the general principles of the operation. As we observed above, the body of the machine serves to house the valves for controlling the fluid and the oil pipes for conveying it to the points at which power is needed. The principal element is the ram and this, with its hydraulic cylinder and piston, is shown in Fig. 172, which is an elevation of the upper portion of the machine.

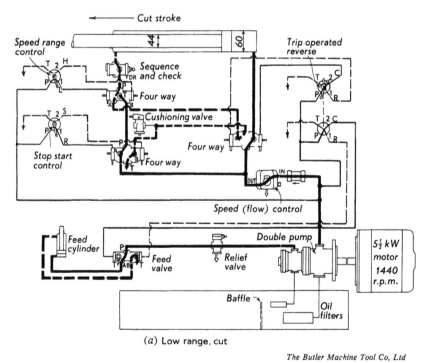

(a) Low range, cut

*The Butler Machine Tool Co, Ltd*

Fig. 173 Circuit diagrams for hydraulic shaping machine

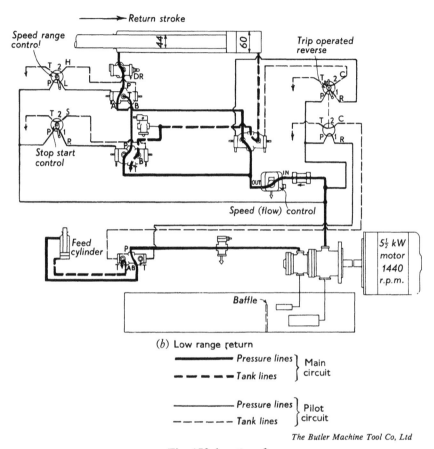

(b) Low range return

————————— Pressure lines ⎫ Main
— — — — — Tank lines ⎬ circuit

————————— Pressure lines ⎫ Pilot
— — — — — — Tank lines ⎬ circuit

*The Butler Machine Tool Co, Ltd*

Fig. 173 (*continued*)

## Further applications of shaping

We have already discussed* some of the more elementary uses of the
shaping machine, so that we may now proceed to consider additional
possibilities of its use. The slide shown at Fig. 174 incorporates features
which occasionally come up for machining. The tee slots and dovetail
slideways may be shaped or milled, but when only one or two components
are in question shaping is probably the most suitable method to use.
This is because milling cutters of the correct size and angle are not always
readily available, and to obtain them specially for occasional use would

* *Workshop Technology*, Part 1, Chap. 12.

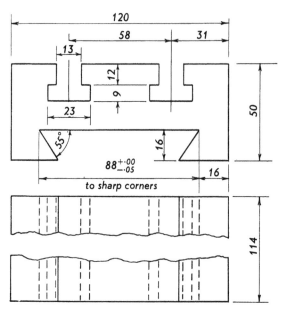

Fig. 174 Top slide. Cast iron. Machine all over

not be in the interests of economy. For the shaper, however, a tool for any special shape can be prepared in a short time, often by grinding the form on to a standard tool which is already near to the form required. A method of dealing with the component in question is as follows:

(Since the slide is a casting, it is likely that a recess will be cast on the side containing the angular dovetail slideways.)

1. Hold in vise. Rough shape top face. Reverse and rough opposite faces. Leave about $\frac{1}{2}$ mm on each face for finishing.

2. Reverse and finish top face. Reverse and finish bottom faces.

During these operations take care that the amount removed from either face allows sufficient to be machined from the bottom of the recess. In operations 2 it will help to achieve parallelism if the job is packed up on parallels to within about 12 mm of the top of the vise jaws and knocked down with a lead or rawhide hammer.

3. Hold in vise. Rough each of the four sides of the block, leaving about $\frac{1}{2}$ mm on each for finishing.

4. Repeat operations 3 and finish the four sides.

In these operations it is important that the sides are square with each other and with the upper face of the slide. The first condition is helped by shaping two opposite sides parallel and then setting one of them up by a square from the machine table. For squareness with the top face ensure that this is clamped evenly against the fixed jaw of the vise. (Operations 3 and 4 could be done by clamping the job to the machine table and shaping down from the head with side tools, setting the block each time by working with a square from the edge of the machine table.)

5. Set the solid vise jaw parallel with the line of the tool motion. Hold the job lengthwise, top face up, and knock down on to parallels, or on to the base of the vise.

Use a parting tool 10 mm to 12 mm wide and take the slots down to depth, spacing them at their correct centre distances. With the same tool open out the slots to about $12\frac{1}{2}$ mm wide (Fig. 175(5)).

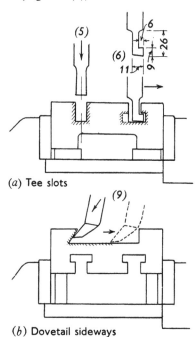

(a) Tee slots

(b) Dovetail sideways

Fig. 175 Shaping operations on slide (numbers refer to operation numbers in text)

6. Prepare R and L hand tools to the form and dimensions shown at Fig. 175(6). Clamp one of these tools in the box, line it up and set it down the slot so that it just scrapes the base of the slot. Clamp up the vertical slides of the machine table and head.

This operation now requires some concentration and nerve, since at the end of each stroke the clapper box must be lifted by hand and held up so that the tool completely clears the job on its return stroke. To allow time to do this the stroke should be set to give an overrun of 75 mm to 100 mm at each end and the machine set to run on a low speed. Failure to lift the tool clear each time will result in a catastrophe! Start the machine and whilst manipulating the clapper box with the left hand put on small amounts of cut (table traverse) with the right hand (Fig. 175(6)). After sinking in the tool to depth on one side of the slot, change the tool for its opposite partner and repeat the process on the other side of the slot. Repeat for the other slot.

7. Open out the upper sides of the tee slots to 13 mm wide. This can be done with the tool used in 6 or a knife tool.

8. Reverse the job in the vise and knock down on to parallel strips. Calculate the *inside* width of the dovetail corners and shape the opening to within about $\frac{1}{2}$ mm of depth and about 1 mm of width, ensuring that the sides of the opening are centrally spaced with respect to the sides of the block. Mark out the angular faces of the dovetail and roughly shape out the metal, using side tools, and remembering to lift the tool clear (as in 6) when the tool is cutting underneath.

9. Grind a pair of R and L hand side tools to an angle of about 45°–50° and take off the extreme sharp point. Set the head of the shaping machine to an angle of 55° with the horizontal, using a vernier protractor (see *Workshop Technology*, Part 1, page 379). Swing the clapper box away from the side to be machined. Clamp one of the tools in the box and set it into the angular groove so that its point will cut whilst its sides are clear. Start up the machine and by feeding from the head take a cut down one of the sloping sides (Fig. 175(9)). Check the angle with a protractor, but it should be correct if the head was set properly. Continue shaping the sloping side until within a few thousandths of the marking out, and at the same time machine the bottom face of the opening as far across as possible, leaving a few hundredths for finishing. Now set the indicating sleeve on the head, so that when it is wound to zero the tool will be set for the finishing cut on the bottom face of the opening. Wind the head back and by a horizontal movement of the table set the tool to finish the sloping dovetail side. Now take the cut down with the head and when the zero mark is reached allow a few strokes of the tool to occur, and then commence to traverse the table so as to feed the tool out of the corner. This will take the finishing cut on the base and when the tool has gone as far as it can be taken the machine can be stopped. (By shaping down the dovetail, and then continuing out

along the bottom face, a cleaner corner is ensured than would be the case if both cuts were taken into the corner.)

10. Reverse the setting of the head and clapper box, change the tool for the opposite hand and commence work on the other dovetail. After machining to within about $\frac{1}{2}$ mm of the marking out, finish machining the base of the opening to blend with the previously machined portion and set the indicating sleeve to zero with the tool in this position. The angular grooves must now be checked with test bars (see page 50) to determine how much remains to be machined off. When this has been found, set the tool to scrape the surface and by using the indicator sleeve of the table traverse take off about three-quarters of the amount remaining. Re-check and if the reading comes out as expected cut may be added to bring the face down to size. If, of course, the reader is too nervous to risk all in the second cut, he may make another check after a further cut and before finishing. For the sake of a clean corner it is a good plan to bring the tool out of the corner by a traverse of the table, and if the tool has been fed to zero on the sleeve this should result in the tool just scraping the previously finished bottom surface.

The job is now complete and only requires the extreme sharpness to be removed from the edges and the corners with a smooth file.

The component shown at Fig. 176 is intended for a setting piece, (see later) and accuracy is particularly important in 1. the angle of the vee and 2. the vee faces intersecting on the centre of the cylinder. For convenience in working it will be best to set the vise jaws at right-angles to the ram and grip the job by its end. For this purpose the ends should be accurately faced and the centre relieved by a few hundredths to a diameter of about 25 mm. This is shown on Fig. 176.

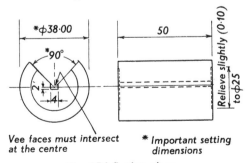

Vee faces must intersect
at the centre

* Important setting
  dimensions

Fig. 176 Setting piece

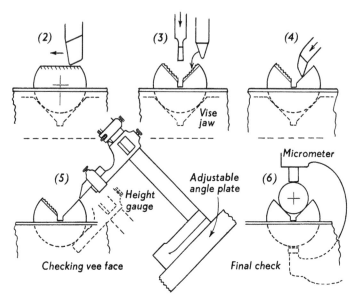

Fig. 177 Operations on setting piece (numbers refer to numbers in text)

Diagrams of the following operations are shown at Fig. 177.

1. Support the job on a vee block and packing to raise its centre slightly more than 2 mm above the jaws. The vee block and packing should not extend for the full length of the job but should leave at least 6 mm clear at each end. Carefully set the job parallel with the table and with the line of stroke.

2. As a guide to the corner extremities of the vee a flat may now be shaped on the cylinder. The width of this will be $2 \times 19 \times 0 \cdot 707 = 26 \cdot 87$ mm, so that if it is shaped to a bare 27 mm, carefully measured with a rule, it will not be far out. (Alternatively, the distance from the flat to the opposite diameter of the job will be $19 + 19 \times 0 \cdot 707 = 32 \cdot 43$ mm.) Shape this flat, checking from the opposite diameter for parallelism.

3. Using a protractor, the vee may now be marked out on one end, and the centre metal roughly taken out. Then with a 4 mm parting tool shape the centre clearance groove to depth.

4. Set the head of the machine to 45° using a vernier protractor or a sine bar to get it correct. Clamp a nice, keen side tool in the box and swing the clapper box upwards. Shape down one side of the vee until within about $\frac{1}{2}$ mm of the edge of the flat.

5. Set an adjustable angle plate to 45° and line it up on the table beside the vise. Using a height gauge from this, measure from the vee to the opposite diameter of the job. This distance, when finished, should be 19 mm, and a check should be made at each end for parallelism. Shape down until the finished size is reached, getting the best possible finish on the final cut.

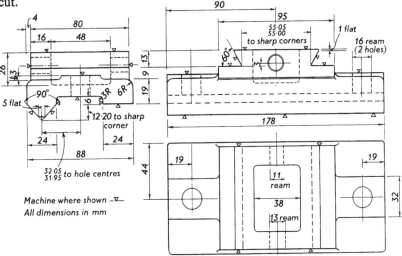

Fig. 178 Machine slide. Cast iron

6. Repeat operations 4 and 5 on the other face of the vee. At a final check a test bar may be placed in the vee and measurements taken (at each end) between this and the opposite diameter of the job. The discussion on page 63 should enable the reader to work out the necessary reading.

The slide shown at Fig. 178 is a suitable type of component for machining on the shaper. Its lower surface has a form which fits with a correspondingly shaped slideway and the upper dovetail tongues carry a slide with ways similar to those on Fig. 174. The slide will be in the form of a casting and the holes will be bored after the shaping operations. Before commencing to machine the component (or any casting) it is advisable to check it over with a rule to ascertain that there is enough material left on the machined faces to ensure that they will clean up. A safer way of doing this, of course, is to mark out the job, but marking out is a rather lengthy and expensive process which is not always justified unless the component in question is very complicated. Furthermore, marking out is only a guide, and should not absolve the machinist from checking and measuring the job as he goes along.

1. Line up the fixed vise jaw with the ram and clamp the job holding on the edges of the base. Place equal packings under each of the end hole facings and knock the component down on to them. Check that the upper surface of the base (now underneath) is reasonably parallel with the machine table and adjust if necessary. Rough shape the base on each of its three sections, using a cut just large enough to remove the scale. The vee can be dealt with by swinging the head of the machine to 45° on each side.

2. Shape the centre, relieved portion of the base, bringing it down so that the base thickness at this section is correct. For this, an angular tool such as that shown at Fig. 177(3) may be used, and the tool can be swung to either side for getting into the corners. As this is a clearance surface, the finish is not important, but the vertical sides of the recess should be carefully faced with a feed of the machine head, as one of them must be used later for setting up.

3. By swinging the head to 45° on either side, and carefully setting with a vernier protractor or sine bar, shape the two sides of the projecting vee. Keep its shape symmetrical with its base and take it down until the distance from its sharp top edge to the surface of the centre portion of the base is according to drawing. Take off the top sharp edge to produce the 5 mm wide flat.

4. The opposite, flat bearing surface must now be finished, and to obtain its correct vertical position with respect to the vee, a setting piece similar to that in Fig. 176 can be used. Place this on the vee and measure to its top diameter with a height gauge from the surface. The surface must now be reduced until this dimension is 19 mm (rad of test piece)+12·20 mm (Fig. 179(b)).

The base is now finished, but before removing the component from the vise carefully determine the vertical distance from the flat bearing surface to the truncated top of the vee. This can be found by setting a pile of slip gauges on the flat surface and checking across with a dial indicator until the heights are equal (Fig. 179(c)).

5. Shape or grind two packing pieces (or one long one) to the thickness just determined, so that with these under the flat of the base the slide will be level. Clamp the casting to the table of the machine and line it up with the stroke of the ram by locating one of the vertical sides of the recessed portion with tenon pieces set in one of the table tee-slots. Clamp a stop against the end of the job. Shape the ends of the dovetail slide.

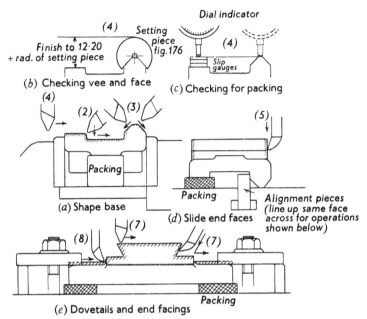

Fig. 179 Shaping operations on slide (numbers refer to numbers in text)

6. Mark out the profile of the dovetail slides on one end. Clamp the casting across the table and carefully set one of the vertical sides of the recessed portion of the base at right-angles to the ram (this can be done by working with a height gauge from the vertical slideways of the machine table, or by using a try-square from the edge of the table) after setting clamp stops behind the job at each end. Using the marking out as a guide, rough shape the top and profile of the dovetailed slides.

7. The top angles and bottom faces of the dovetail slide may now be finished and checked, as explained in connection with the example of Fig. 175(9). As in that example, after finishing one side and base to the marking out, take the opposite one down in its base, and check the dovetail dimension with rollers for a guide as to finishing. In this case the two horizontal bottom faces must be *in the same plane*. Before finishing the second one, therefore, check with feeler gauges under the tool on the first face, and then adjust the cut so that the faces are both at exactly the same height.

8. With the casting still in the same position, and the clamps moved to clear the tool, machine the end facings for the two 16 mm holes. The

setting piece (Fig. 176) would again be useful for setting to locate the boring of the two 16 mm holes.

### The dividing head on the shaping machine

Although the dividing head is most usually associated with the milling machine, there are rare occasions when it may be usefully employed on the shaper. This is chiefly because a shaping tool can cut right up to an obstruction which would be fouled by the milling cutter. For work held in a chuck the ordinary milling machine head may be bolted directly on to the table of the shaping machine, but when the work necessitates a tailstock the increased length involved requires an auxiliary base-plate. It is not general, however, to employ the milling machine head for shaping, but to use a simplified pattern. This is because the jobs involved generally require less complicated increments of division and, as the equipment is not in regular use, it would be uneconomic to provide costly equipment to spend most of its life in the stores. An example of a shaping dividing head is

*The Crowthorn Engineering Co, Ltd*

Fig. 180 Shaping machine dividing head

shown at Fig. 180. Indexing is performed through a steel worm engaging a 40T wormwheel and an index plate having nine rows of holes on each side. The holes provided are such that all numbers up to 50, all even numbers from 50 to 100 and a range of numbers beyond this are possible. The spindle is bored for a No. 3 Morse taper and screwed to accommodate a chuck.

### Dividing head work

The ratchet teeth on the wheel shown at Fig. 181 may be conveniently cut on the shaping machine, since a milling cutter would run into the gear

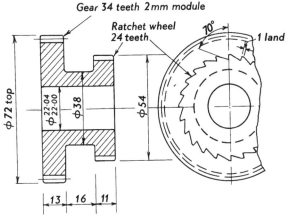

Fig. 181 Ratchet and gear wheel

before it had cleared its cut. The gear teeth would, of course, be cut on a milling machine. A convenient method of holding this would be to use a small stub arbor of the type shown at Fig. 155(e), the head being used in the horizontal position. Failing this the gear may be put on a mandril and for the sake of rigidity, one end of this could be held in the chuck of the head, with a collar threaded on the mandril to take the thrust of the cut against the chuck jaws (Fig. 182). After the head and its base have been lined up on the machine, setting lines for 24 teeth should be scribed on the blank, as explained in connection with Fig. 155(b). The teeth are cut with a parting tool having an angle of 20° (90 − 70) ground on its edge and the cutting procedure is similar to that we have discussed for the milling example on p. 201 except that care will be necessary to feed the tool down at a rate at which it can cope with the cut.

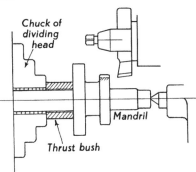

Fig. 182 Set-up for shaping ratchet teeth

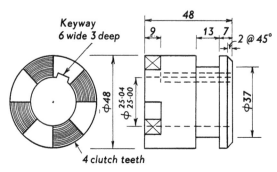

Fig. 183 Dog clutch sliding sleeve

Another job suitable for shaping is that of cutting the teeth on the dog clutch shown at Fig. 183. This could not be milled with a disc type cutter, arbor mounted, since any cutter over about 40 mm diameter would run into the opposite side of the metal before it had cleared its cut. The dogs could be cut with an end mill, although the size necessary to do the job would be rather frail. The arrangement for doing the job on a shaper is shown at Fig. 184, and the mode of procedure as an inset at (b). According to the rigidity of the dividing head it might be necessary to rough out the cutting with a parting tool about 4 mm wide, but the procedure for finishing the dogs is as follows:

Using a parting tool about 6 mm to 8 mm wide, set the RH corner (from m/c front) on the centre, and shape a slot to full depth, as shown at Fig.

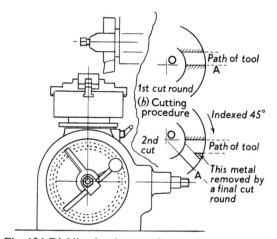

Fig. 184 Dividing head set up for shaping clutch teeth

184(b) (i). Index the work through successive divisions of 90° and repeat the operation until four such slots have been completed. Now move the job over until the opposite corner of the tool is on the centre, index the job in a clockwise direction through 45° and take this cut down four times as before, Fig. 184(b) (ii). The spaces will now be left with a small remaining piece of metal in the centre, which can be removed by four subsequent indexings and cuts.

### The planing machine

The work on the planing machine is held on the machine table and moves under the tool which is carried on slides called the rail, supported in the form of a bridge over the table ways. Broadly speaking, the scope of the machining operations on planing and shaping machines is similar, the difference being in the dimensions of work which may be accommodated. Although shapers are made with strokes up to 1 metre, the most usual maximum is about $\frac{1}{2}$ to $\frac{3}{4}$ metre, and work longer than this is best placed on the planer, since increased length is usually accompanied by size increases in other directions and the large, solid table of a planer is a better

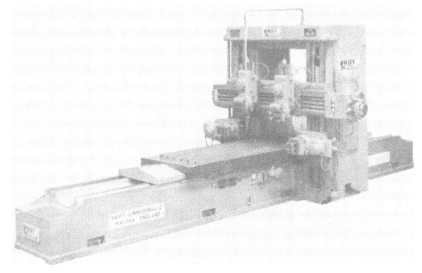

*George Swift & Sons, Ltd*

Fig. 185 Planing machine, 3 m × 1·25 m

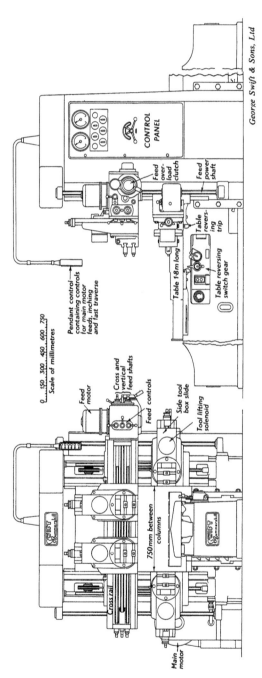

Pendant control
containing controls
for main motor
feeds, inching
and fast traverse

0  150  300  450  600  750
Scale of millimetres

Feed
over-
load
clutch

Feed
power
shaft

CONTROL
PANEL

Table 1·8 m long

Table
rever-
sing
trip

Table reversing switch gear

Feed
motor

Cross and
vertical
feed shafts

Feed controls

Side tool
box slide

Tool lifting
solenoid

750mm between
columns

Cross rail

Main
motor

*George Swift & Sons, Ltd*

Fig. 186 General arrangement of 1·8 m × 0·75 m planing machine

means of support for heavy work. The size of a planing machine is specified by the length of its stroke and the width which will pass between the columns supporting the rail. The bed, of course, is approximately twice as long as the table, since it must provide support for it when either end is below the tool situated above the centre of the bed.

An illustration of a 3 m × 1·25 m planer is shown at Fig. 185 and a general arrangement of a 1·8 m × 0·75 m machine at Fig. 186. On the latter diagram the principal elements have been labelled.

### Table drive

The longitudinal motion of the table is obtained by driving through a pinion or worm on to a rack attached to the length of the underside of the table. There are two general methods of driving the rack: (1) through a train of gears to the pinion engaging the rack, or (2) by means of a single shaft and a worm which engages the rack. The second method necessitates that the worm shaft should enter the planer at an angle and we show at Fig. 187 a diagrammatic sketch of the arrangement. The variable speed and reversing features on the main driving motor are obtained by feeding it with a variable voltage direct current of reversible polarity from a special motor-generator set which forms part of the equipment of the machine. Electrical driving facilitates the arrangement of tripping and reversing, the trips actuating switches which make the necessary electrical changeover. This method gives a light, quiet and responsive set of conditions and is free from

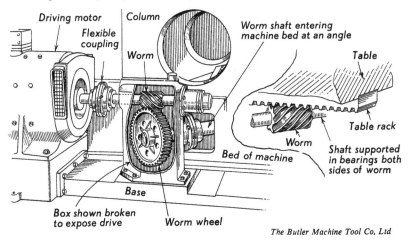

*The Butler Machine Tool Co, Ltd*

Fig. 187 Diagrammatic sketch of table drive of 'Butler' spiral electric planer

the drawbacks, associated with mechanical means of doing the same job. The tripping arrangements vary with different makes of machine. On the Swift machines we have illustrated, the trip cams are carried on the table itself and are set to actuate the switch-gear through a rocking lever, the arrangement of which can be seen on Figs. 185 and 186. The 'Stirk' machine employs the principle of attaching the trips to a circular flange whose revolving movement synchronises with the longitudinal motion of the table. This is shown at A on Fig. 188.

*John Stirk & Sons, Ltd*

Fig. 188 Floor-mounted feed motor and controls of planing machine

## Tools and tool support

Cutting tools for the planing machine are in general similar in their nose shapes to those used on the lathe and shaping machine. Owing to the heavier work they have to perform, however, they are made from material of larger section to give them a strong rigid shank. Thus, whilst the average lathe or shaper tool has a shank about 26 mm × 20 mm, that for a planer would be about 32 mm square. The additional width results in the plan profile of the nose having a blunter angle than is the case for the narrower tools. The tool is carried in a box after the style of the shaper, and this box is carried on a vertical slide which, in its turn, is incorporated with horizontal slides for fitting to the slides of the rail. The tool, therefore, may be fed vertically on its slide and horizontally across the rail. The manual operation for both these movements is controlled by shaft ends which project from the end of the cross-rail. It is usual, on all but the smallest machines, to provide a pair of tool box slides on the cross-rail, and a further pair of side tool boxes (one each side) sliding on the columns. These latter provide a very convenient means of planing vertical surfaces. The vertical travel of a side head is not limited as is that of the others, and it is conveniently placed both with respect to the operator and to the position that faces to be machined are likely to occupy on the table (i.e. near its edge). Increased accuracy is likely to be obtained since the ways on which this box slides are large and permanently aligned in a position perpendicular to the table surface. Most planing machines are provided with a positive means of raising the tool to clear the surface on the return stroke of the table, instead of relying on the tool to drag across the work and lift the clapper box. The force to do this is often applied by an electric solenoid.

### Feeding

We have already shown a mechanical method of feeding planer tools (Fig. 67). On the all-electric machines the feeding movements are operated electrically by a separate electric motor and control box. These may be mounted at the side of the machine as at Fig. 188, B and C, or on the machine itself as shown and designated in Fig. 186. In addition to the provision of a normal range of cutting feeds on all slides, these facilities provide for controlled approach of the table to the tool ('inching') and for rapid power traverse. Electric machines can also be provided with many other refinements such as electrical means for locking up slides, motor elevation of the cross-rail, electrical interlocking safety arrangements, etc.

### Table and work support

The work-supporting principle of the planer is very good since the table is supported by solid metal right down to the foundations of the machine. This feature, which is shared by some other machines such as the slotter and borer, is an important one from the aspect of rigidity and freedom from deflection and vibration. On machines such as the shaper and knee type milling machine the table is carried in such a manner that it overhangs from vertical slides, and although supports and legs may be provided there cannot be the rigidity associated with a correctly designed method of solid support.

### Work setting

For the most part, the work done on the planer is held by being bolted or secured direct to the table, with the help of such simple and fundamental methods of support as may be devised by the operator. It is true that holding equipment and fixtures are used, but their application is limited to special work in quantity. The vise is not much used, since if a job is small enough to be held in this way it is of a convenient size for the shaping machine and should not be wastefully taking up capacity greater than is necessary. Sometimes, for planing the edge of a long strip, two vises may be used in line for holding the work after its thickness has been machined up.

To set, support and clamp work to a flat table provided only with tee slots, often demands much skill and ingenuity, and for a planer employed on a general run of work the setting up forms no small part of the skill associated with operating the machine. The usual tackle a planer possesses comprises bolts, nuts, washers, clamps (straps), packing, wedges, angle plates, vee blocks, jacks, parallel strips and anything else which might be useful. In time a good selection of such accessories are collected, as it is never known what the next job may necessitate in equipment for setting up. Setting up and clamping a job in this way demands, as we have remarked, considerable skill and experience. The following are the chief problems:

1. The work should be clamped and supported so that it will neither move nor deflect when subjected to the forces of the cut. (This is more important in planing, where the full force of the cut is applied suddenly at the beginning of each cutting stroke, than in other machines where the pressure is continuous when once started.)

2. The clamping should not introduce distortion of any kind. Some distortion, when relieved by unclamping, will result in machined surfaces becoming inaccurate. Distortion is generally avoided by ensuring that there is adequate support under each clamping point.

3. On early operations the work should be set and tested to ensure that the first machined surfaces, which will be used as datum planes for the remainder of the machining, are left in a suitable relationship to allow the subsequent surfaces to clean up. (If the job is marked out it will only be necessary to set up to the marking, but marking out is an expensive operation, and in many cases is unnecessary if machinists know their job.)

**Examples of planer work**

*Example 4. To plane up the strip shown at Fig. 189(a).*

1. Pack and clamp the strip to the machine table with a stop to take the thrust of the cut. Set the table trips to permit a suitable overrun at each end and take a cut over one face of the strip (Fig. 189(b)). For jobs such as this where normal clamping would interfere with the tool, a method of clamping such as we have sketched may be used. The clamping lugs fit into the tee slots and clamp the work through the spikes. These must be pointed and hardened at the end in contact with the work and must be set in a slightly inclined position so as to exercise a pulling down tendency on the work. The number and spacing of the clamping points will depend on the job, usual practice being 150 mm to 300 mm centres. Distortion must be avoided by placing thin packings under any substantial portions of the surface not in contact with the table, so that the work feels solid, and cannot be rocked in any direction.

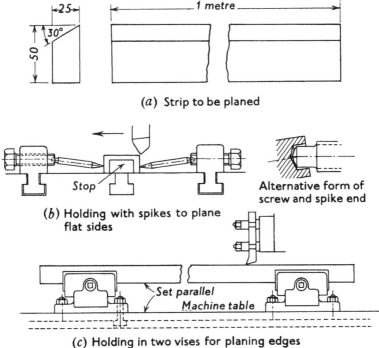

(*a*) Strip to be planed

Alternative form of
screw and spike end

(*b*) Holding with spikes to plane
flat sides

(*c*) Holding in two vises for planing edges

Fig. 189 Settings for planing strip

2. Turn the strip over on to the machined face, clamp as before and finish off to thickness.

3. Hold the strip between the machined faces, using two vises, set the under edge parallel with the table and clean up one edge.

4. Line up the solid jaws of both vises so that they are in line and parallel with the travel of the tool as explained for the shaping machine. Hold the strip between its flat faces and set the previously machined edge parallel with the table (use parallel strips or height gauge). Set the head of the machine to the correct angle and machine the angle until there is sufficient land to test with a protractor. Test, and correct if necessary, and continue machining until the width is correct (Fig. 189(c)).

*Example 5. To machine the bed casting shown at Fig. 190(a).*

### 1st setting

1. Set the casting on its feet on the machine table. Feel for any rock and if present pack until the bed rests solid on its four corners. Run the point of the scribing block over the top surface to check that it is reasonably parallel with the table. Line up the sides of the casting from the tee slots, so that they are parallel with the travel of the table. Clamp down and put stops at the end.

2. Take a roughing cut across the top surface and repeat if necessary until the surface is within about 1 mm of size (Fig. 190(b)).

3. Swing over the head to 35° from the vertical and rough down the dovetailed edge; continue until there is about 1 mm on the surface when measured from the centre of the casting (Fig. 190(c)). (When doing this operation the tool must be lifted clear of the face by hand on the return stroke or it will dig in.)

4. Reset the head vertical and rough face the side facings, using the same tool as in 3 (Fig. 190(d)).

5 and 6. Repeat operations 3 and 4 on the other side of the casting, leaving about 1 mm on both surfaces.

The remainder of the work at this setting consists of finishing the surfaces roughed in 2 to 6 above. On the first side finished, the sharp corner of the dovetail should be measured from the centre of the bed. Before finally finishing, the dovetail should be checked for angle and the side facing for squareness. When the second side is finished the overall width to sharp corners may be measured. The final operation at this setting consists of removing the sharp edges of the dovetails with a straight tool.

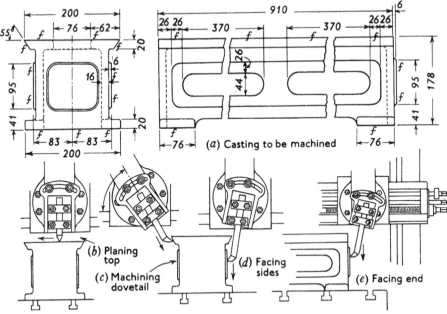

Fig. 190 Settings for a planing bed casting

## 2nd setting

Turn the casting over and strap down, making sure the surfaces of the table and casting are clean.

Rough and finish the surfaces of the feet.

## 3rd setting

Slew the casting round, set and clamp it so that the side machined facings are parallel with the cross-rail slides. This setting may be made with a scribing block, or with the tool and feeler gauges. Rough and finish the end facing using a side tool and the vertical travel (Fig. 190(e)). (The reader should note that operations 4, 6 and the 3rd setting could be performed with the side heads on machines so provided. In the 4th and 6th operations the cutting could then proceed simultaneously with the planing of the top surface.)

## Miscellaneous set-ups on the planer

We show at Figs. 191 and 192 various types of work being machined on planing machines.

*The Butler Machine Tool Co, Ltd*

Fig. 191

The operation at Fig. 191 is that of taking the finishing cut over the face of a casting. The component is set and clamped on to packing to clear projecting metal underneath. The side head of a machine is shown working at 192(a), facing the edge of a machine bed. The complete machining at this setting is as follows:

(1) Top faces machined with the two cross-slide boxes; (2) inner edges machined using vertical feed of top toxes; (3) outer edges faced with side boxes; (4) under faces of outer edges. The underside of a grinding machine tailstock is being machined at (b) and the operation in progress is that of cutting the V where the base fits the dovetail.

The machines shown at Fig. 191 and 192(b) are of the 'openside' type where the rail, instead of being supported at both ends, is carried at one end and overhangs. This feature renders the machine more useful for certain classes of work where the restriction of vertical pillars at both sides of the table would prevent the job from passing through.

**(a)**

Fig. 192.     (b)

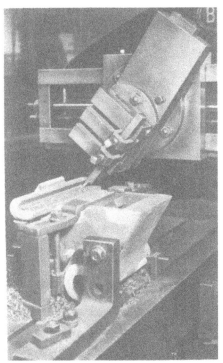

## The slotting machine

The slotting machine is equivalent to a vertical shaper and, as such, is capable of undertaking a range of work not conveniently held and machined on the shaper. A diagram of a precision slotting machine with a maximum stroke of 200 mm is shown at Fig. 193, from which its chief features can be seen. This machine is available with strokes of 200 mm and 300 mm and other production type machines are made with strokes ranging from 355 mm to just over 1 m. On account of the nature of the work they normally undertake, slotters do not need such a long ram travel as shapers and a general purpose machine could do most of its work with a maximum stroke of about 200 mm to 350 mm.

*The Butler Machine Tool Co, Ltd*

Fig. 193 Precision slotting machine (200 mm stroke)

The driving arrangement for the ram of the machine shown is illustrated on the cut-out section of the machine at Fig. 194. The drive from the motor is transmitted by enclosed vee belts to a four-speed gearbox through a clutch with a built-in brake. The large helical toothed stroke wheel shown in the diagram takes its drive from the gearbox and this wheel carries a bronze sliding die block, which can also be seen. This block slides in ways cut on

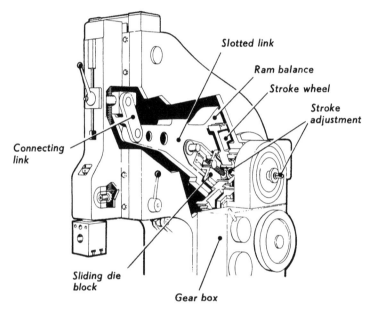

*The Butler Machine Tool Co, Ltd*

Fig. 194 Driving arrangement for ram of slotter

the face of the large oscillating 'slotted link' which is connected to the ram of the machine by means of the forked connecting link shown. The end of the slotted link opposite the machine ram is enlarged to balance the ram. The stroke of the machine is adjusted by varying the radius at which the sliding die block rotates with the stroke wheel and this is effected by means of the square-ended shaft which moves the block via the bevel gears and screw, all of which can be identified on the diagram. The vertical positioning of the ram is effected from its front through another pair of bevels and a screwed connection to the fastening at the forked link, which can also be seen on the diagram. Readers familiar with the slotted link shaping machine mechanism, will observe a certain similarity between the two.

A feature of the slotter adding greatly to its usefulness is the circular table which is rotated by the handwheel at its front. The shaft to which this wheel is attached carries a worm which meshes with a wormwheel fixed to and mounted underneath the table. In addition to hand-operation the circular table may be given an automatic feed for the purpose of traversing a cut. The use of this table is very convenient for machining circular shapes; in fact, it is almost the only method available for obtaining many of the shapes required in toolmaking practice. The angular graduations also render the table a valuable help on angular work and where divisions of the circle are required. To assist in divisional accuracy, the table of the machine shown at Fig. 193 is provided with an index plate similar to a milling dividing head. For normal rectangular working the circular table may be locked and table movements obtained by means of the longitudinal and cross-slides, both of which may be traversed by automatic feed.

## Cutting tools for slotting

The slotting tool is supported, and moves relative to the work as shown at Fig. 195(a). This differs from other applications of cutting with a single-point tool insomuch that, instead of cutting on a line perpendicular to its length, it cuts parallel with its shank. To allow for such changed conditions

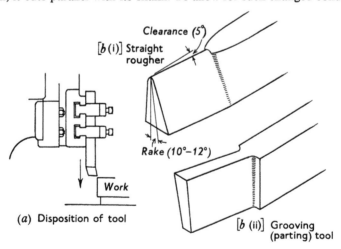

Clearance (5°)

[b(i)] Straight rougher

Rake (10°–12°)

(a) Disposition of tool

Work

[b (ii)] Grooving (parting) tool

Fig. 195 Slotter tools

the tool shape is considerably modified from that associated with other single-point tools, and it resembles a half-round chisel. As we have previously discussed (Part I, p. 167), the action of the usual tool bears little similarity to our usual conception of cutting, but that of the slotter tool bears still less resemblance, seeming almost to 'push' the metal off the face being machined. Fig. 195(b) shows two examples of slotter tools, that at (i) being a straight, rougher type, whilst the tool at (ii) is used for getting out slots, keyways and corners. Slotter tools have front rake, no side rake, and front and side clearance, the approximate values of these angles being given on the sketch. The cutting angles are not appreciably altered for cutting different materials. The shank needs to be of a large and rigid section, as the tendency in cutting is to deflect the tool away from the work, an action depending for its resistance upon the strength of the tool and the rigidity of its support.

**Setting up**

To allow for the tool to overrun the surface being machined and still stop short of the table, all work must be supported on packing to raise its lower surface a small distance from the table (about 12 mm is sufficient). The main precautions to be observed are that the packing is parallel, supports the work as near as possible to the cut, and under the clamping points. During the course of machining round a profile it may be necessary to move the clamps and packing to allow the tool to complete its work without fouling. This may necessitate careful manipulation if important settings are not to be disturbed.

For setting lines and surfaces parallel to the axes of the cross and longitudinal slides the table tee slots may be used after the table has been rotated and set to zero on the angular scale; it should always be locked, of course, when not being used in rotation. Keyways, slots, etc., in round work should preferably be marked out together with centre lines, so that these may be used for setting to ensure that the slot axis bisects the centre of the work. Circular profiles should be marked out together with their centre if this happens to be located on the work. By sticking a pin to the tool with a piece of chewing gum, disengaging the worm and rotating the table by hand, the job may be set in position by adjusting until the pin point follows the marked profile, or seeing that the centre dot (if centre is marked) is concentric with the pin point.

**Examples of slotting machine work**

*Example 7. To slot the keyway in the boss shown at Fig. 196(a).*

1. Scribe the centre line across the boss and mark the width of the keyway. Set the job on a pair of parallels about 10 mm to 12 mm high and clamp down. Rotate the table until the marked centre line is parallel with the cross-slide of the machine (test with a pin).

2. Obtain a keywaying tool about 10 mm wide and clamp it in the tool holder. Adjust the stroke of the ram to suit and move the ram so that the tool just clears the table at the bottom of the stroke.

3. Set the tool central to the marking out and commence to slot the keyway, feeding very gradually by hand. When nearing the depth check with inside calipers from bottom of keyway to opposite side of bore. This should be 56 mm when finished.

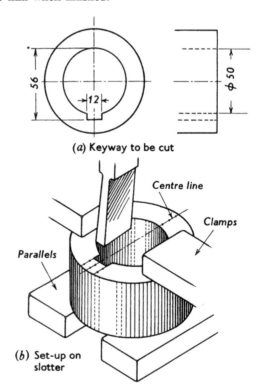

(a) Keyway to be cut

(b) Set-up on slotter

Fig. 196 Slotting a keyway

4. When within about ½ mm of depth wind away from tool, move side-ways until edge of tool will slot one side of keyway to the marking out. Finish that side to the marking out using a very fine feed.

5. Obtain a 12 mm plug or slip gauge and finish the other side of the keyway until it is a tight fit to the gauge. Finish the bottom of the keyway Fig. 196(b)).

*Example* 8. *To machine the hole in the block shown at Fig. 197(a).*

1. The block will have been milled or shaped to its outside sizes. Mark off the shape of the hole, mark and drill holes as shown at Fig. 197(b).

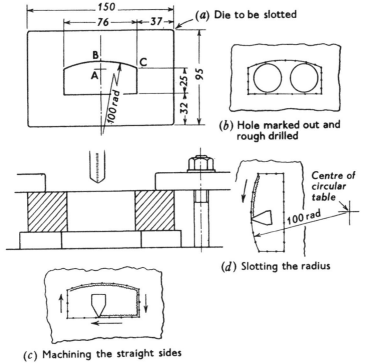

(a) Die to be slotted

(b) Hole marked out and rough drilled

(d) Slotting the radius

(c) Machining the straight sides

Fig. 197 Operations for slotting hole in die

2. Pack on parallels and clamp to machine table, setting out at 100 mm radius, and adjusting until the marked radius portion is true against a point attached to the tool. Rotate the table until the long straight side of the slot is parallel with the longitudinal slides of the machine. Note the angular reading and lock the circular table.

3. Set a roughing tool in the tool box and rough away the surplus metal by hand feeding until the profile is to within about 1 mm of the marking out.

4. Put a reasonably sharp-nosed tool into the box, and using cross feed for the short sides and longitudinal feed for the length of the hole, take the straight sides of the profile almost to the line. It will be necessary to rotate the tool half-way through this operation in order to get right round.

5. With the same tool and circular feed of the table, machine the rounded portion almost to the line.

6. Reset the table to the first position, touch up the tool and finish the straight surfaces as explained in 4 (Fig. 197(c)).

7. Finish the rounded portion (Fig. 197(d)).

(*Note.*—As a check, the dimension AB may be calculated as follows: From the intersecting chords of a circle, $AB(200-AB) = AC^2$ i.e. $200AB - AB^2 = 38^2$. This reduces to the equation $AB^2 - 200AB + 1444 = 0$ from which $AB = 7.5$ mm) (Fig. 197(a).)

8. With a pointed tool, slot out the corners sharp.


*Example* 9. *To machine the cam lever shown at Fig.* 198(a).

Before coming to the slotter, this lever would require to have its boss bored and faced, and the sides of the curved foot machined.

These operations could be carried out on the lathe:

(*a*) Chuck on boss, bore 25 mm, face one side of boss, face one side of curved foot.

(*b*) Hold on mandril between centres, or on peg turned up to suit from material held in chuck. Face other side of boss and curved foot.

The centre line and sides of the keyway and the centre and straight portion of the foot profile should now be marked out at one setting of the block as shown at Fig. 198(b). To mark the curved profile accurately would involve placing the point of the dividers on a level with the sides of the foot. This could be effected by transferring the centre to packing attached to the lever but is hardly necessary as the slotting can be done by setting up from the centre point.

1. Set the lever on parallels or on a bush, and slot the keyway as discussed in Example 7.

2. Bolt the lever down to face of boss opposite to marked centre point of foot profile. (Bolt through hole.) Set the lever from a pin attached to the tool until the centre point of the radius is concentric. Place 6 mm pack-

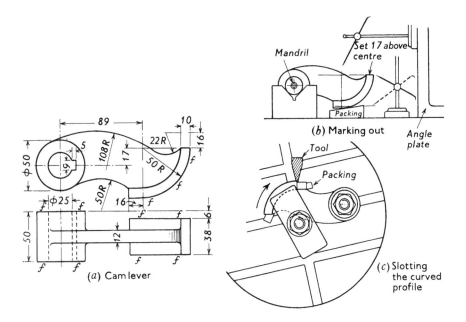

(b) Marking out

(a) Cam lever

(c) Slotting the curved profile

Fig. 198 Operations for machining cam lever

ing under the foot side, and put a clamp on top of the foot keeping the end of the clamp clear of the tool when it passes round the profile.

3. Set the straight portion of the profile in line with one of the machine axes and note the reading of the circular table.

4. Commence to slot the straight portion and when the commencement of the curve is reached stop traversing along but take up the traverse on the circular table.

5. Continue slotting until down to size (Fig. 198(c)).

*Example* 10. *To machine the hexagonal hole in the bush shown at Fig. 199(a).*

As it comes for slotting the bush will be turned and faced to size, and bored about 36 mm to 37 mm. It may be marked out but if not, this must be done; plug the hole, find the centre point, scribe a circle to the corners of the hexagon (43·88 mm) and step round with the dividers. This gives the corners, and the sides may be marked by joining up.

1. Set the bush on a thin ring having a hole slightly greater than the corner dimension, or on two parallels. Hold through the centre with a

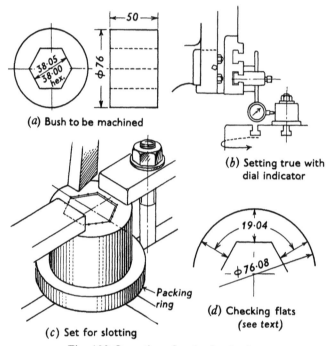

(a) Bush to be machined

(b) Setting true with dial indicator

(c) Set for slotting

(d) Checking flats (see text)

Fig. 199 Operations for slotting bush

washer and set the outside true using a dial indicator held in the tool box (Fig. 199(b)).

2. Put two or three clamps round the outside and take away the central clamping bolt.

3. Rotate the table until two sides of the hexagon are in line with the longitudinal slide, take the reading and lock up.

4. Put in a straight (parting) tool and slot to within about $\frac{1}{2}$ mm of the marking out. Add 60° 0′ to the table reading, index round and do the same for the next side of the hexagon. Repeat until the six sides are all roughed out (Fig. 199(c)). (When indexing take care about the direction of the backlash (lost motion) on the table-operating wheel shaft. For example: if the last movement of the initial setting is made with an anti-clockwise rotation of the wheel, this direction should be preserved throughout. If during indexing the mark is overshot, do not just reverse to the mark but come back a full turn and regain the mark with a movement in the original direction.)

5. The hole must now be finished concentric with the outside of the

bush. Measure the outside diameter. Index round to a flat which is clear of the clamps, and measure with a micrometer between the flat and the outside diameter. When the flat is finished this must be $\frac{1}{2}$ (out. dia. $-$ 38). If the outside diameter is, say, 76·08 mm, this dimension will be $\frac{38 \cdot 08}{2} = 19 \cdot 04$ mm (Fig. 199(d)). Finish slotting the flat until this dimension is attained and take the reading of the cross-slide indicating sleeve.

6. Index round to each position and finish slotting each flat to the same reading on the cross-slide sleeve. (This must be done as in some positions the clamps will prevent the micrometer being used for checking.)

7. If necessary put in another tool with sharp corners to clean out the corners.

**Machining internal grooves**

The slotting machine with circular table provides a convenient means of cutting internal splines, castellations, etc. Fig. 200 shows two examples of such work, and it would be carried out in the same manner as the internal hexagon discussed in Example 10. Unless a sized tool is used, after setting a suitable tool on the centre as for Example 8, no longitudinal movement would be necessary except for opening out the keyways to size. For the castellations shown at (b) a vee tool would be used, and it would be fed to the same depth each indexing. The indexing of the circular table would have to be carried out to suit the number of slots being cut, the angle being found by dividing 360° by this number. For internal grooves of this type slotting is the only convenient method for the production of one or two articles, as the alternative is broaching or drifting, and unless a reasonably large

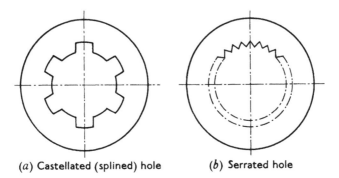

(*a*) Castellated (splined) hole      (*b*) Serrated hole

Fig. 200 Work suitable for slotting when in small quantities

order were concerned it would not be economic to make the necessary special equipment.

When cutting internal splines it should be ascertained whether the top or the bottom diameter of the spline fits and locates on the mating shaft. If the top diameter, then the keyways must have their bottoms to the correct radius and concentric or the wheel will not run true. If, on the other hand, the inner bore forms the fitting location, the stopping point for the slotting tool is not so important.

External teeth and grooves may also be cut in this manner on the slotter, but when a suitable milling cutter is available milling is a more convenient method for such work.

# 8  Drilling and boring

The production of cylindrical holes is such an important part of work-shop practice that numerous types of drilling and boring machines are available, each being adapted to certain classes of work. Holes may be produced by drilling with twist drill, flat drill, D bit, etc., or by boring with one or more single-point tools or cutters mounted in a bar or head. Reaming is a finishing process which may be applied to drilled or bored holes and is generally used to give a better finished and more accurate cylinder than the originating method which preceded it. Reaming will not correct any previous positional inaccuracy in the hole, since the reamer follows the axis of the hole it is finishing. The choice between drilling, drilling and reaming, boring, or boring and reaming, depends on several factors, but chiefly on the size of the hole, the accuracy of the hole (particularly as regards its position) and the number of holes to be produced.

For example, if one or two components required a 32 mm hole, accurately positioned, they would probably be bored with a single-point tool, but if a large number of such components were involved they would be drilled and reamed, a special jig being designed to hold and position the component, having bushes for guiding the drill and reamer to the correct position.

When holes are larger than about 38 mm to 50 mm diameter, boring, or boring followed by reaming are the most common methods of produc-tion. This is generally because most components in which there are holes greater than 50 mm are in the form of castings, with the hole cored out about 6 mm to 9 mm less than finished size. Boring is the most accurate and convenient method of producing the hole from such a condition, as a two-flute twist drill digs in with its corners and will not cut, and a three or four-flute drill tends to follow the cored hole, which can never be true with its finished form. Boring is frequently followed by reaming to the finished size, because this method of finishing enables a size to be attained much more quickly and consistently than is possible with single-point boring tools.

## Drills and accessories

In addition to the two-lipped twist drill with which we have already dealt,* there are several additional types of drills and accessories likely to be of interest at this further stage.

**Straight flute drill.** The rake which the helix angle of the flutes imparts to the cutting edge of a twist drill is apt to be a disadvantage when drilling brass, or thin sheet metal, as the lips tend to dig in. For such purposes and any others where cutting rake is unnecessary a straight flute drill may be used, since the straight flute gives a cutting edge on which there is no rake (Fig. 201(a)).

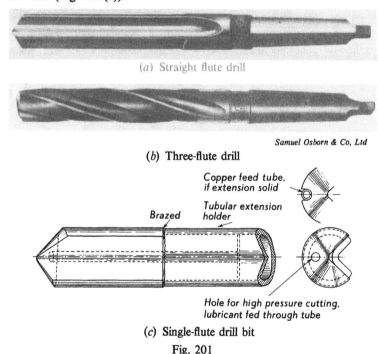

(a) Straight flute drill

*Samuel Osborn & Co, Ltd*

(b) Three-flute drill

Copper feed tube, if extension solid

Tubular extension holder

Brazed

Hole for high pressure cutting, lubricant fed through tube

(c) Single-flute drill bit

Fig. 201

**Three and four-flute drills.** When a hole already exists and has to be opened out to a diameter only a small amount larger, the two-flute drill is not suitable as its corners tend to dig in and render the operation almost impossible. Such cases occur when opening out cored holes in castings and when, due to insufficient power, large holes are drilled out in two or

* *Workshop Technology,* Part I, p. 265.

three stages. The three or four-flute drill enables work of this nature to be undertaken successfully, the multi-lipped point centring itself in the previous hole and the cutting, being shared by more cutting edges, proceeds without trouble. These drills will not drill from the solid (Fig. 201(b)).

**Single-flute drills.** For drilling deep holes a drill bit of the shape shown at Fig. 201(c) may be used. According to the length of hole, this is often attached to a shank, or brazed into a tube, the cross-sectional shape of which is a continuation of the flute, thus allowing the swarf and chips to get away. Running down to the drill point is a copper tube let into a groove milled in the body of the drill. Arrangements are made for this to be connected to a high-pressure supply of cutting lubricant which is ejected in considerable volume on to the region of cutting, cools the drill and in its escape along the flute of the drill carries away the swarf and prevents the drill point from becoming clogged. Drills of this type are not rotated but are used on machines on which the work is rotated, the drill being held stationary.

**Straight-shank drill chucks.** For holding drills up to about 14 mm diameter, which are commonly made with parallel shanks, a drill chuck is necessary. These are often call 'Jacob's' chucks, but this name has only been given to them because the make of the same name is so well known that the term is often applied to chucks of this type in general, and irrespective of their make. The construction of one of these will be gathered from Fig. 202, which shows an outside illustration and section of the 'Cardinal'

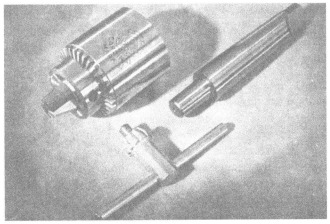

(*a*) Outside view of chuck
Fig. 202 Drill chuck

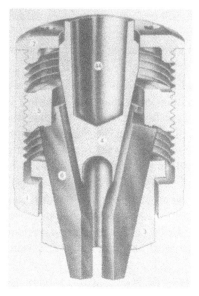

Brooke Tool Manufacturing Co, Ltd
(b) Section

Fig. 202 (cont.) Drill chuck

chuck. The internal thread of the operating sleeve (1) moves the jaw carrier (5) when the sleeve is rotated. The jaw carrier holds the jaws (6) in bayonet slots and the jaws are thrust forwards at an angle determined by the grooves in the body (4) and the coned bush (3) through which they travel. To increase the turning effort gear teeth are provided on the end of the sleeve and these engage with a key which pilots in radial holes in the coned bush. To obtain the maximum concentricity and grip the key should always be used in each of the three hole positions. A short taper hole in the top of the chuck enables it to be accommodated to a taper shank for fitting to the machine spindle.

**Quick-change holders.** The operation of changing one drill for another may occupy more time than that taken with drilling the hole, and when several different holes have to be drilled in the same component quick-change holders are often used. These enable one drill to be substituted for another in a few seconds, if necessary, without stopping the machine. A diagram showing the components of one of these holders is shown at Fig. 203. Each drill or tool to be used is fitted into one of the interchangeable collets and the holder is fitted to the machine spindle. To change the tool whilst the machine is running the sleeve (A) is grasped and moved upwards. This allows the retaining segments (B) to move clear of the collet, which can be withdrawn. Both the sleeve and the knurled collar on the collet are free and cease to rotate when grasped in the hand. The drive is obtained by the tang at the top of the collet engaging a slotted driver in the chuck.

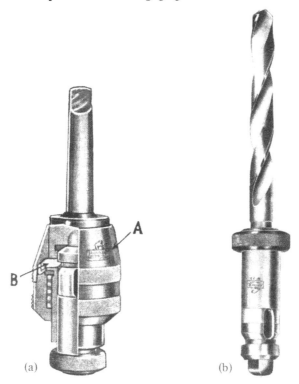

*Frank Guylee & Son, Ltd*

Fig. 203 Quick-change collet

**Using drills with broken tangs**

The angle of the taper forming the shank of a drill is sufficiently fine for the frictional grip between the shank and the hole to provide enough torque to drive the drill. If, however, one of the members becomes damaged, or a small particle of foreign matter is present when they are assembled, the advantage of the fit will not be obtained and all the driving force will be thrown on the tang. This may cause the tang to be twisted off, and render the shank useless in an ordinary taper hole. In many cases when this occurs the drill is otherwise in good condition, and to avoid the necessity of discarding it a special holder may be used. One type of such holder is shown at Fig. 204, in which it will be seen that the taper hole is made to

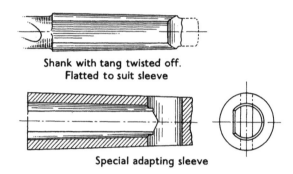

Shank with tang twisted off.
Flatted to suit sleeve

Special adapting sleeve

Fig. 204 Method of accommodating taper shank with broken tang

accommodate a taper on which a flat has been machined. The damaged shank may thus be used by machining a flat on it until it fits the hole in the collet.

## Drilling machines

Machines for drilling take a variety of forms, depending upon the size and class of work for which they are to be used. In addition, a selection of machines of the same type may not look very much alike owing to variations in design, driving arrangements, etc. It will only be possible for us, therefore, to enumerate and explain the most general standard types, giving illustrations which we consider to be representative in each case.

**The vertical machine.** In the construction of this machine the work-table and drill spindle are carried on a vertical column. There are two general varieties of construction: (1) column of box section, and (2) column of round section. The box column arrangement (sometimes called the *upright machine*) has the work-table incorporated with a bracket which slides on ways at the front of the machine. Support and elevating movement for the table is provided by a telescopic screw underneath its centre. The drill spindle is carried in a bearing bracket which is clamped to slides above those which carry the table. This bracket is only moved on its slides during the process of adjusting the drill to a suitable distance from the table; the movement for feeding the drill is separate and incorporated in the bracket which carries the spindle. When its slides are clamped this machine is rigid, and because of the short overhang of the drill a stiff unit is formed capable of efficient duty (Fig. 205). For high production purposes machines of this type are often equipped with multi-drilling heads similar to that shown at

*Fredk. Town & Sons, Ltd*

Fig. 205 Vertical drill (box column, 750 mm)

Fig. 219, or with a turret similar to that shown at Fig. 229 but provided with drilling equipment.

The round column type (often called the *pillar drill*) follows the same general design as that above, the principal difference being that the table, instead of being supported on vertical slides, is carried on the round pillar (Fig. 206). This makes for some versatility, since for high jobs the table may be swung to one side and the work supported on the base of the machine. At the same time the mode of support for the table is less rigid than that of the upright drill and places a restriction on its width.

Upright and pillar drills are usually specified according to their 'swing', this being twice the distance from the spindle axis to the nearest face of the machine body or pillar. Other important details are the maximum diameter

*Fredk. Pollard & Co, Ltd*

Fig. 206 Vertical drill (530 mm capacity—round column)

of drill that can be used on steel, the maximum height that can be accommodated with the table in its lowest position and the length of spindle feed. The speeds and feeds on a drilling machine will have been chosen by the maker as including the most suitable range for accommodating the range of drill sizes for which the machine is designed. A table of twist drill feeds is given on p. 188 W.T. Part I.

**The radial drill.** On this machine the drill spindle is carried on a slide which is supported on ways machined along a radial arm. This arm is supported at its end by a pillar and is capable of angular rotation about the pillar as centre. The drill spindle, therefore, because of its angular and radial movement, is able to cover the entire working surface of the table and its precincts, a property which renders this machine much more versatile than the upright type of machine with its fixed spindle. Work in which a number of holes are to be drilled may be set up and clamped to the radial machine, and all the drilling carried out by adjusting the drill to the positions of the various holes. On the upright pattern the work would have to be moved for each hole. An additional factor in favour of the radial machine for operation on a general assortment of work is its greater capacity for large jobs. A 600 mm upright or pillar drill measures 300 mm from the spindle to the nearest point on the machine, so that it would not be possible to deal with a hole more than 300 mm inside any surrounding metal. Comparable with such a machine in drilling capacity would be a 1000 mm radial machine. This would allow the drill spindle to operate, if necessary, at a distance of 1000 mm from the centre of the column and, allowing for the column, would admit about 900 mm between the drill and the nearest point of the machine.

The work supporting arrangements on radial drills vary. On some types the base is a flat bed at floor level with an auxiliary box table which may be used when necessary. Other machines have the column built from a box table at some height from the floor, whilst another type (often of the lighter size) has a table supported from the round column after the style of the machine shown at Fig. 206, except that the table is rectangular and projects as far as the end of the radial arm. A radial drill, with an auxiliary box-type table, is shown at Fig. 207, and the reader should observe and study other types as well as the work carrying methods on other machines with which he may come into contact.

**The sensitive drill.** A sensitive drilling machine is one designed essentially for using small drills (up to about 10 mm to 14 mm). The sensitive element

*Adcock–Shipley*

Fig. 207 A radial drilling machine (914 mm)

about it is in the lightness and delicacy of its control which humours small, weak drills, and if the machine is being fed by hand, enables the progress of the drill to be felt, so that when the drill gets into trouble the pressure on it may be relaxed. All the varieties of machines above are made in the sensitive pattern, so that we may have sensitive upright, sensitive pillar or sensitive radial machines. In addition there are numerous designs of sensitive bench drilling machines which are made to fit on a bench or table. Some of these are high-speed machines made to operate only with very small drills (e.g. a 1 mm drill to cut at 18 metres per minute should run at about 5750 rev/min).

**Grouped spindle machines.** Drilling machines of the upright and pillar type are often made with two, four or six spindles placed fairly close together and served by a common table. For some classes of work a unit of this type is advantageous and economises in time. For example, if a number

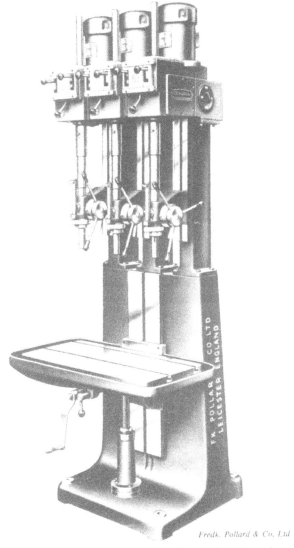

Fig. 208 Three-spindle column unit sensitive drill (330 mm)

of jobs require three or four operations involving various sized drills, or drills followed by other tools such as taps or reamers, each of the respective spindles may be fitted with a tool and the work completed by passing along from one spindle to the next. By this means drill changing, which often takes longer than the cutting operation, is eliminated. Group machines of this type are also often worked after the fashion of a number of single machines with an operator to each spindle. Under such conditions each operator may work independently, or a progressive system may be operated as we have just discussed. A three-spindle unit of the upright type is shown at Fig. 208.

### Driving of drills

In the machines illustrated we have tried to include machines with various methods of drive in order to convey an impression of different practices. A common method with belt-driven machines in the past was to take the drive to a shaft or gearbox situated near the floor, and then take it up again to the drill spindle. Why this design persisted for so many years was a mystery, as the obvious alternative was to strengthen the machine frame and then take the drive to somewhere higher up. From the aspect of safety alone such a change was worth while, the long protruding belt being a source of danger as well as an eyesore, and the guards necessary to enclose it spoiling any effort at cleaning up the floor space and the appearance of a shop. Fortunately, independent motor drive has cleaned up the appearance of drilling machines in this respect, since the motor may be incorporated and blended into the design at the most convenient spot.

### General purpose drilling

When used in a general purpose machine shop or toolroom, the drilling machine is called upon to carry out a variety of work for all and sundry. For large work the most suitable type of machine is a radial, as sooner or later something will require drilling which cannot be accommodated between the drill and the column of an upright machine. As a second choice the pillar machine is the most versatile, because the table may be swung aside to allow high work to be supported on the base. For the smaller range of work sensitive machines are necessary. In the writer's experience there has often been a gap between the sensitive and heavy machine, with no suitable machine available for using the range of drills between about 10 mm and 20 mm, a common range of holes in most shops. The speeds of sensitive machines are generally too high for holes above 9 mm, and the larger machines do not lend themselves to the most efficient use of drills in the range mentioned. Machines are made for covering this range efficiently, but probably when sensitive and the larger types have been installed the middle range is forgotten.

**Marking out.** In drilling, more than in any other operation, the work should be marked out before commencing. The accuracy with which this should be done depends, of course, on the work. Sometimes a rough marking out with a centre dot to locate the start of the drilling is all that is necessary, whilst for more important work the centre lines will have to be obtained accurately, and the hole circles scribed and dot punched to provide a guide for the drilling. The limits of accuracy expected from drilling are not fine, as it is not a precision type of operation, but even so, it should be possible to work to a tolerance of about $\frac{1}{2}$ mm when necessary. In many cases it is more important to be able to drill holes in two separate components whose centres will match up when the components are put together. This property is important for stud, bolt and dowel holes in components which must be fastened together. When doing such work it is advisable to mark out and drill one component and then use it as a template from which to drill the other. If, as is generally the case, one hole is clearance for a bolt, whilst the other is a smaller, tapping hole, the choice of two methods of working is given. Either the parts may be clamped together, the tapping hole drilled right through and the holes in the one part being opened out afterwards, or the part with the larger holes may be marked out and drilled first, and then used as a template for countersinking the other hole positions with the drill just used. The first method is probably the more reliable, as the chances of a drill following the axis of a smaller drilled hole are better than its picking up an axis from a countersunk spot.

**Holding the work**

This subject has been referred to earlier in this book, but it is of sufficient importance to mention again. Many people when using the drilling machine either seem too impatient to spend time clamping the job properly or else look upon drilling as something rough and inferior, and therefore hardly worth the trouble of setting up with thoroughness. The result often is broken drills or cut fingers, and if a milling or shaping job is worth proper clamping, the same applies to drilling. For dealing with the average range of work on the drilling machine small and medium machine vises (the low, flat type), bolts, clamps and packing, vee blocks, parallel strips and angle plates will be necessary. One of the angle plates available should be of the adjustable type shown at Fig. 191(a), *Workshop Technology*, Part I, as such a plate permits work to be set up for drilling angular holes which would present difficulties without such help.

### Drills and cutters

A comprehensive selection of drills should be available, and even when an English size may be required the wide selection of metric sizes available will often provide a drill to suit the case, particularly if a metric equivalent chart is hung near the machine or stores. Reamers should be provided for nominal sizes, and counterbores, facing bars and cutters, countersinks, etc., should be available in as wide a variety as possible. A good selection of reducing collets and one or two Jacob's chucks, well maintained, will enable the most efficient service to be obtained from the machine and tools.

### Examples of drilling

Various examples of drilling work are shown at Fig. 209. At (a) is shown work being done on the end shield of an electric motor. The holes being drilled are at an angle, and this is being obtained by tilting the table of a radial machine. The operation consists of first drilling a 3 mm hole for a depth of 19 mm followed by a 4 mm hole 10 mm deep. The final operation is that of spot-facing 13 mm diameter, and the job illustrates the advantage of quick-change collets, two being seen ready for fitting. Note that the machine is a light radial type with the work-table tilted to the angle of the job. At (b) is a locomotive cross-head being drilled for the tapping holes to take the liner screws. These holes are subsequently tapped on the machine. The operation at (c) is that of spot-facing a previously drilled boss on a pump housing. Note the method of lining up the work by setting it on two pins passing through previously drilled holes. The jigs for dealing with the drilled holes can also be seen on the machine table. The operation at (d) is interesting as it serves to illustrate the application of a two-spindle machine. The operation is drilling and tapping inlet and outlet holes in gear pump bodies, and the components are mounted on fixtures which support and locate in the recesses for the gears. The operator loads a casting under the drill spindle, lowers the drill and engages the feed. He then moves over and loads a drilled component under the tapping spindle and starts the tap. By this time the drilling operation is complete, and another casting is loaded

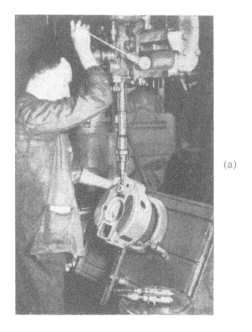

Fig. 209 Miscellaneous drilling operations (see text)

James Archdale & Co, Ltd

(c)

Alfred Herbert, Ltd

(d)

Fig. 209 (*continued*)

whilst tapping is proceeding. The drill spindle is provided with a stop, and the tapping spindle trips and reverses when the tapping is completed.

## Drilling jigs

When a quantity of similar components have to be drilled marking out becomes uneconomic and impracticable, since holes drilled carefully to marking out will not necessarily be interchangeable. The use of a jig not only eliminates marking out, but also introduces precision into the process as the jig is provided with bushes which guide the drill and enable holes to be drilled to close tolerances as regards centre distance. In addition, inter-changeability is assured between components drilled in the same jig. Drill-ing jigs vary widely in design. They may hold the component or support and locate it, or be in some form of plate which is fastened to the work. In its design the jig incorporates means for positioning the work, or itself to the work, so that the bushes provided for guiding the drill are located in the correct relationship with other surfaces concerned. These bushes are of steel, hardened and ground with their bore made to suit the drill and their outside diameter a press fit in holes bored in the body of the jig. If reaming is to follow drilling two bushes are used which are a push fit in a larger, permanent bush fastened in the jig. After drilling has been performed through the drill bush this is replaced by the reamer bush for guiding the reamer. The use of jigs, as well as fulfilling the functions we have men-tioned, is advantageous for the following reasons:

1. They can be designed to provide efficient holding and supporting arrangements for awkwardly-shaped components and save elaborate setting-up.

2. They reduce setting-up time and give increased output. (Without the use of jigs, the huge outputs of interchangeable parts, now so com-mon, would not be possible.)

3. Their use reduces costs as, in addition to the saving in time, a lesser skilled and lower paid operative may be employed where they are used.

The design and the amount of capital to be expended on a jig will depend, of course, on the scale of production. Even for only a few components it might be advantageous to make one of a simple, cheap type in order to save marking-out time and ensure that the holes drilled are interchange-able. The chief points in connection with the construction of drill jigs are:

1. The jig should be as simple as possible and not over ambitious in its scope. For example: rather than drill four 6 mm holes and one 38 mm hole in the same jig it might be better to split the work between two jigs,

as a sensitive machine would drill the small holes and a heavier machine the 38 mm hole.

2. Guide bushes should be as long as possible, and their lower end should terminate near to the surface being drilled. Short bushes wear quickly and provide little directional guidance to the drill.

3. Where locating and supporting surfaces are liable to wear they should be made so as to be renewable. If there is any risk that the component might not be contacting a location, the position should be made visible if possible.

4. The jig should be as light as possible (for handling), and unless it is to be used permanently clamped down, its feet should lie outside the drilling points.

5. It should be easy to load the component into the jig and the arrangement should be as foolproof as possible.

In general, owing to the complicated shape necessary, the body of a jig is made as a casting with drill bushes and locations of case-hardened steel or, if small, of hardened high carbon steel. Clamps, clamping screws, etc., are usually made of mild steel and hardened locally at the points of contact. By slight modifications in the design it is often possible to construct a jig body from mild steel plating and rolled sections cut to shape and welded up to the desired form. This construction is lighter than a casting, and there is less danger of a fracture if the jig is dropped. In addition the expense of a pattern for the casting is saved.

Sketches of drill jigs with the components they accommodate are shown at Figs. 210–12. The jig at Fig. 210 is for drilling six holes round the rim of the component shown. The job fits and locates on the central peg and is clamped with the slotted washer, a feature which permits the

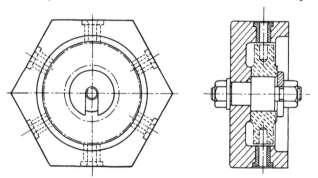

Fig. 210 Jig for drilling holes in rim of disc

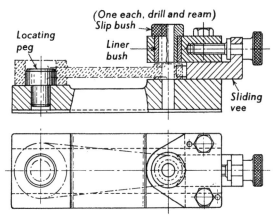

Fig. 211 Jig for drilling and reaming small end of lever

Above: Component to be drilled.
Below: Built-up drill jig for operation.
Location faces are marked $f$.
Screws etc. holding jig together
not shown

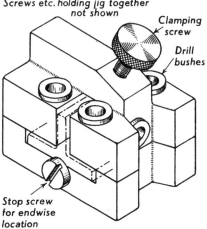

Fig. 212 Drill jig

component to be loaded and unloaded without screwing the nut right off. Drilling is carried out by standing the jig on each of its six faces and drilling through the opposite bush. The radial and spacing accuracy of the holes is controlled by the flats, and the boring for the bushes.

Fig. 211 shows a jig for drilling and reaming a hole in the end of a lever, locating from the previously finished hole in the opposite end. The lever is slipped over the locating plug and its end swung under the drill bush where the rounded end is centralised for drilling by a sliding vee. The centre distance is controlled by the accuracy of the centres of the plug and drill bush. The diagram shows the slip bush, one each of which is fitted for drilling and reaming respectively.

*Alfred Herbert, Ltd*

Fig. 213.

For drilling the T-section bracket shown at Fig. 212 the jig shown alongside would be suitable, locating and clamping arrangements being indicated on the sketch.

Fig. 213 shows the drilling of 14 oil holes in a piston. This particular operation is automatic once the jig has been loaded and the cycle set in operation, the spindle of the machine being cam operated and the indexing

(a)

Fig. 214 Examples of
drilling with jigs

(b)

*James Archdale & Co, Ltd*

of the piston controlled by compressed air. The operation at Fig. 214(a) is that of recessing a hole with a special single-blade eccentric recessing tool. The component is a pneumatic pick handle. The reaming of a 35 mm hole is being carried out at Fig. 214(b), the component being shown in the foreground. In the jig shown the following work is done: drill and counterbore four 13 mm holes, drill eight M8 tapping holes, drill and spot-face one M10×1 tapping hole, drill and ream one 35 mm hole.

### Facing and counter boring

These operations may be carried out on the drilling machine either with the tools we have already described, or with special ones made to suit particular conditions. A common and useful type of facing tool consists of a flat cutter held in a bar as shown at Fig. 2, p. 7. The best conditions are obtained if the leading end of the bar is a running fit in the hole whose boss is being faced, but if the bar is short and stiff this is not necessary. If such a bar is used on work held in a jig the bore of the liner bush must be large enough to admit the cutter, and if the height of the facing is important a collar may be pinned or held to the bar with a set-screw in such a position that it stops the cutter when facing has proceeded to the required level (Fig. 215(a)). Counterboring may be carried out and its depth controlled in the same way, but if the counterbore has no pilot on its end to take a location in the hole its outside diameter should be a fit in the bush (b).

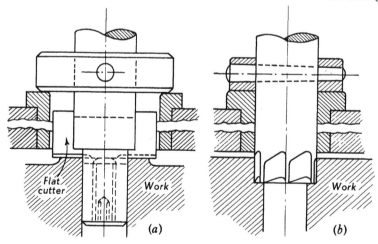

Fig. 215 Facing and counterboring in jig, with stop collar to control depth

## Drilling deep holes

The drilling of a deep hole occurs occasionally in the experience of most workshop engineers, and the subject often provides material for the reminiscences of the old timers, comparable with the best fishing or golfing stories! How many of us have not heard the story of the flat drill on the end of tubes fed through in lengths after the style of chimney sweeps' rods, and emerging at the other end as true as a die?

A deep hole is usually classified as one whose length is more than five times its diameter, and the difficulties attendant on the drilling of such holes are as follows:

1. The drill has a tendency to run out, a fault which when once started grows worse as the drilling proceeds.

2. A difficulty is encountered in getting the chips away from the front of the drill to prevent choking up and seizure, and promote free cutting.

3. It is difficult to lead sufficient cutting fluid to the cutting point for the purpose of keeping it cool and lubricating the action.

The faults, in a sense, tend to aggravate each other, as the type of drill which is the most free cutting and the easiest to which cutting fluid may be applied (e.g. the flat drill) is the worst type for preserving its concentricity because of the absence of any backing support in the previously drilled portion of the hole.

The factors which influence condition 1 are as follows:

(a) *The relative movement of work and drill.* At first sight it would appear immaterial whether the work or the drill revolved to obtain the relative rotation for cutting, but for deep drilling it is easier to produce concentric holes when the work is revolving, with the drill held stationary. This may be explained by Fig. 216. At (a) the hole is being drilled by a revolving drill which has taken the inclined direction AB. Having once taken this direction the drill obeys an elementary law of mechanics and follows it, and the deeper it goes the more will it deviate from the true axis AC. As the hole gets deeper also, the length of drill which fits in the drilled hole, backing up and guiding the point, is easily able to overcome the flexual tendency of the drill to right itself, a tendency which the drill cannot take advantage of in any case since it only cuts on its point. The conditions when the work is revolving are shown at (b). In this case, if the drill point deviates from the axis of rotation of the work it is not allowed to follow any set direction, but is rotated in a circle, having a radius equal to the amount of deviation. This is shown as *r* in the diagram which, by the dotted lines,

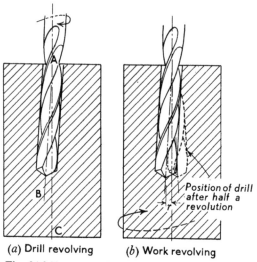

(a) Drill revolving   (b) Work revolving

Fig. 216 Trueness of drilling long holes (see text)

shows the position of the drill when the work has made half a revolution. Under these conditions, therefore, the drill point tends to follow the path of least resistance and remains on the axis.

(b) *Starting the hole.* A great help when a long hole has to be drilled in the lathe is to get a good, true start with it. This can often be done by drilling with a slightly smaller drill for a depth of about two diameters, and then boring this out to the diameter of the drill to be used. The bored hole, being true, will help the drill to remain axial during the early stages of its work, and if this can be accomplished the rest is generally straightforward.

(c) *Quality of material.* It is useless trying to drill a true hole in a material which is not homogeneous such as wrought iron. A break in the continuity of the material such as would be caused by a streak of slag in wrought iron is sufficient to cause the drill to deviate from its true direction. Blow holes in cast iron may also lead to a similar effect.

The problems of clearing the swarf from the drill point and cooling the drill are to some extent similar. When a twist drill is used, unless the drilling is carried out by feeding the drill from underneath so that the chips fall away by gravity, the drill must be withdrawn from time to time. This permits the drill to be cleared and washed with cutting fluid. When drilling from above a magnetised round file is useful for withdrawing ferrous swarf from the bottom of the hole. For the systematic drilling of long holes, however, the twist drill is not satisfactory, and other types are used which are more free cutting and which lend themselves more readily to being flushed with lubricant. We have already discussed one of these (Fig. 201(c)). Another type is the flat drill secured to the end of a bar or tube with suitable arrangements for conveying a high-pressure supply of soluble oil. Owing to the absence of any form of body to back this drill from behind, the shank or tube carrying it is often provided with a number of loose collars having an outside diameter to fit in the hole. As the drilling proceeds these are pushed in the hole to follow up, support and guide the shank in the proper direction. A diagram showing a flat drill fixed to a bar for deep drilling is shown at Fig. 217.

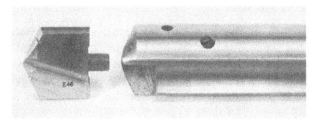

*Alfred Herbert, Ltd*

Fig. 217 Boring with flat drill (note tube sunk into upper portion of bar for delivery of cutting fluid)

For the reasons we have already discussed, long holes are mostly produced by rotating the work and maintaining the drill stationary. For many types of work extreme accuracy is demanded and obtained, as the reader will appreciate when he considers such operations as boring out rifle barrels and jobs of a similar nature. This work is highly specialised and is carried out on machines designed for the purpose.

**Trepanning**

This is a process by which a cylinder is turned by feeding a hollow cutter over its length, the turned diameter entering the bore of the cutter. The trepanning cutter has teeth on its end only, the radial width of which must be sufficient to cover the metal to be turned away. The process may be carried out on the lathe, drilling machine, or any machine on which the work and cutter may be held, and the necessary rotational and feeding motions obtained. It is generally limited to the production of comparatively short diameters, and to cases where alternative turning methods are not convenient. Unless special arrangements are made regarding the cutter,

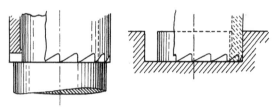

Fig. 218  Trepanning cutter

the length that may be trepanned is limited to the distance from the face of the teeth to the bottom of their bore. A diagram of trepanning is shown at Fig. 218.

**Special drilling machines**

In a book such as this it is only our intention to deal with standard machines, but the drilling machine has been so highly developed for special work that the fact is worthy of passing notice. For certain types of production, of which the automobile industry is a good example, large numbers of components are required in which there are a multiplicity of holes. In order to drill as many of these as possible in the minimum of time, various designs of multi-drilling machines have been evolved. Some of such machines are entirely special, being made expressly for the particular job and suitable for none other. These machines often have several heads, each holding a number of drills which feed from several directions at right-angles, and so complete in a short time the drilling of several perpendicular faces. A more common type is the cluster type multi-drilling machine, which drills holes in the horizontal plane only. This has a number of spindles, each with its own driving shaft and attached to the head of the machine. Within certain limits these spindles are adjustable, so that such a machine is adaptable to a variety of work. A flat jig plate is necessary, of course, to guide the drills into their correct positions and, if the holes being drilled are blind, the entire set of drills must be carefully adjusted as regards length. A close-up photograph of a multi-drill at work is shown at Fig. 219.

## Horizontal boring

The horizontal boring machine is used for boring holes in and machining facings on work where the holes are too large to be dealt with on a drilling machine, and the shape of work not suitable for handling on a lathe or vertical boring mill. In its modern version the machine is an extremely

Fig. 219 Cluster type multi-drill

versatile tool, and when suitably equipped is capable of facing with single-point tool, screwcutting, and for short lengths, external turning. The constructional details of the machine are shown at Fig. 220, in which it will be seen that the bed is provided with long slideways on which is mounted the saddle, carrying the table moving on slides perpendicular to the bed ways. At the left-hand end of the bed is a column carrying the spindle head which moves on vertical slides for adjustment to varying heights of

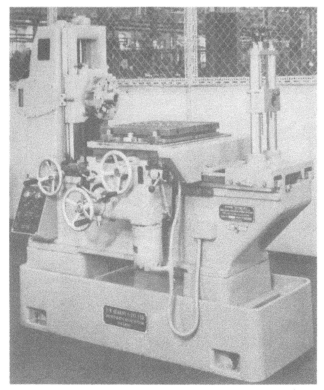

*Staveley Machine Tools Ltd, Kearns-Richardson Division*

Fig. 220 Horizontal boring and facing machine

work. At the opposite end of the bed is a standard bracket which carries a bearing for supporting the end of the boring bar. The spindle is in two parts, composed of a central core similar to a drilling machine spindle which can be fed longitudinally, and an outer spindle flanged at its working end for carrying the facing equipment. Driving is accomplished by taking the power from a motor mounted low down at the back of the column which drives a nine-speed gearbox by means of a toothed belt. From here the drive is taken up inside the column to the spindle slide by means of a flat belt composed of a combination of plastic and chrome leather. It is claimed that this method minimises vibration on the spindle, and reduces the risk of distortion by heat. Power-driven feeds are usually provided for the spindle, longitudinally for the saddle, and crosswise for the table. Plain boring, therefore, may be traversed either by feeding the table with the spindle locked, or vice versa. Milling with cutters carried on an arbor is rarely carried out on the machine, but face and end milling are common, as by such means faces may be machined at the same setting, and hence perpendicular with the axes of bored holes. When these operations are carried out the saddle and spindle are locked and the cut traversed by means of the cross movement of the table. Facing may also be performed with a flat cutter similar to Fig. 2, or by making use of a head attached to the boring bar, which incorporates a slide and actuating mechanism for feeding the tool radially.

*Staveley Machine Tools Ltd, Kearns-Richardson Division*

Fig. 221 Facing head of boring machine

An important additional feature is a facing head which makes facing with single-point tools a simple matter. This is constructed on the end of the spindle, and as shown at Fig. 221 consists of a flange provided with a diametral slideway on which moves the tool-carrying bracket. Radial feeding arrangements are provided for as well as means for locating and locking the bracket at the centre when the machine is used for boring with a long bar supported at both ends. The assistance of this feature adds much to the scope of the machine when working with unsupported stub-boring bars, making possible the fine adjustment of tools for facing, and when the tool is turned inwards, external turning.

**Tools and cutters**

Boring operations on long bores, and on those situated at some distance from the spindle of the machine, are carried out with a long boring bar upon which the tools are mounted. The bar is held and driven by the machine spindle, and its outer end takes a bearing in the tall bracket located at the end of the bed (see Fig. 220). The tools and cutters used may be single-point tools carried directly in the bar, or in a head attached to the bar, or some form of solid tool such as a shell or rose reamer pinned to the bar. The disadvantage of boring with a tool held in a concentric bar is that additional cut for enlarging the hole can only be applied by moving the tool. This involves loosening and disturbing its securing arrangement, and unless a setting gauge is used the amount of movement cannot be determined until a sufficient length of cut has been taken to allow a measurement to be made. This may be avoided by mounting the tool in a head which incorporates means for radially adjusting the tool (Fig. 222). Fixed diameter cutters such as permanent tool heads, shell and rose reamers, etc., are used often for finishing operations on standard holes,

*Staveley Machine Tools Ltd, Kearns-Richards Division*

(a) Three roughing and one adjustable
finishing tool

(b) Two tools

Fig. 222 Boring heads

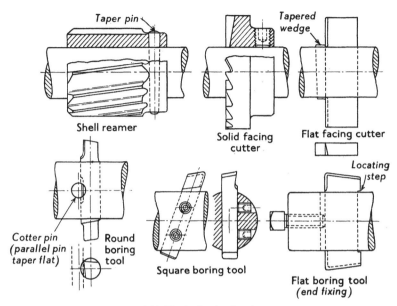

Fig. 223 Tools for boring bars

when the volume of similar work justifies the cost of making them specially for the job. A selection of sketches showing various bar tools and their securing methods is shown at Fig. 223. A very useful tool for short bores and other operations close to the machine spindle is the snout toolholder shown at Fig. 224(a). By employing various tools and settings, this fitment may be used for facing and turning as well as boring, and an example of its applications is shown at (b). For boring small holes at high speed a head of the type shown at Fig. 168 is available. Screw adjustment is provided for the tool and the graduations of movement read to 0·02 mm.

**Setting-up**

Not being a primary operation, boring is carried out on work which has already been subjected to previous machining processes; generally, the finishing of the requisite flat surfaces have undergone shaping, planing or milling. It is important that when the work is set up these surfaces should be correctly aligned so that the axes of the holes to be bored will bear correct relationships to the surfaces concerned. The machine table is the chief agent upon which settings will need to be based, advantage being taken of the properties that the spindle axis is parallel with the table surface,

*Staveley Machine Tools Ltd, Kearns-Richardson Division*

(a)

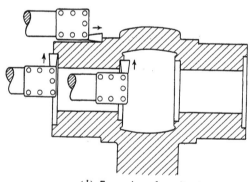

(*b*) Examples of application

Fig. 224 The snout boring tool

and perpendicular with its edge or with the axes of its tee slots. Thus the axis of a bored hole will be parallel with a surface clamped in contact with the table, and perpendicular with a face set both square with the table surface, and parallel with its side or tee slots. When the work is set up these alignments of the surfaces concerned will need to be made, using square, depth and height gauges, dial indicator, and such other instruments as are the most expedient for any particular case. When it is necessary for a hole to be bored at a particular distance from the base, or from a parallel side face of the casting, a taper shank test bar may be inserted in the spindle and working from this with the height gauge the spindle may be set at the correct position. Vertical settings may be made direct from the table surface and horizontal measurements from the surface of an angle plate clamped to the table and set with its face square with the table edge.

*Staveley Machine Tools Ltd, Kearns-Richards Division*

Fig. 225 Auxiliary square table-top for boring machine

A feature which is of great value in promoting the correctness of alignments, and in saving the time of re-setting for different operations, is the auxiliary table-top. This is shown at Fig. 225, and is pivoted on a central spigot with locating and locking arrangements provided, so that it may be located and secured at any of its 90° positions, and also set and locked at any other angle. By the use of this, holes may be bored and faces machined in four planes at right-angles without moving the work, and, after what we have said regarding the value of doing the maximum amount of machining without moving the job, the reader will appreciate the importance of the feature in the promotion of accuracy.

**Horizontal boring set-ups**

It is not possible, in the space at our disposal, to convey a complete impression of the scope of the horizontal borer, but we give at Fig 226 a selection of examples of its application. These will serve to show the versatility of this machine.

(a) Line boring.

*Staveley Machine Tools Ltd, Kearns-Richardson Division*

(b) Face milling.

Fig. 226 Operations on horizontal boring machine

(c) Outside turning.

Fig. 226 (*continued*)

## Boring fixtures

Production work on the horizontal boring machine is assisted by the employment of boring fixtures which incorporate means to locate the work, with brackets for supporting and guiding the boring bars to give the correct hole centres. A simple example of such a fixture is shown at Fig. 227(b), in which the component (a) to be bored with two holes in each end and faced is located on the base with guide brackets at each end and in the centre. The brackets are bored to the correct centres and bushed to be a running fit for each boring bar, these and the cutters being specially made for the job. In operation the component would be clamped to the fixture base (previously lined up on the machine) and located, either from previous machining or from the casting profile. The boring bars would be threaded through the brackets and component (cored holes), at the same time the shell reamers and stop collars being threaded loosely on the bars at the position they will be required. The operations would then be as follows:

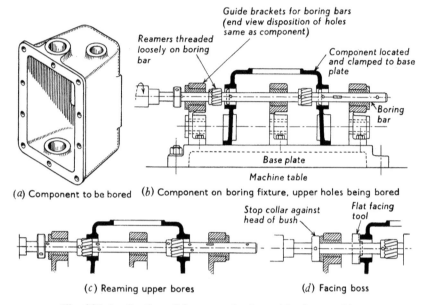

(a) Component to be bored   (b) Component on boring fixture, upper holes being bored

(c) Reaming upper bores          (d) Facing boss

Fig. 227 Application of fixture on horizontal boring machine

1. Fit the rough boring tools to the upper bar and couple its end to the machining spindle. (To assist in connecting bars of this type to the spindle, instead of a taper shank they are often provided with some simple method of coupling them to an adaptor fitting in the spindle.)

2. Rough bore both upper holes (Fig. 227(b)).

3. Remove tools and secure shell reamers with taper pins. Ream holes (c).

4. Disconnect reamers, fit facing tool and stop collar at LH end. Face LH boss (d).

5. Face RH boss in the same way after fitting facing tool at RH end and pinning on its stop collar.

The remainder of the job consists of treating the lower hole in the same way.

### The vertical boring mill

This machine is not well named, as in its application it fulfils more the function of a vertical lathe with permanent chuck and faceplate, than that of a boring machine. The constructional details of a vertical boring mill are shown at Fig. 228. At the front of the machine is the horizontal re-

*Webster & Bennett Ltd*

Fig. 228 Vertical boring mill

volving table for accommodating the work. This can be fitted with chuck-jaws for circular work, or the tee slots may be used with bolts and clamps for setting up and holding irregular work. A horizontal cross-rail, somewhat similar to that on a planer, is supported on vertical slideways and carries the toolholder slide or slides. The machine shown is the 1·22 metre size (table diameter) and is provided with a toolholder in the form of a turret. This enables five tools to be accommodated and economises in tool chang-ing and setting on work requiring several operations. Alternative designs of machines are available with one or two tool slides provided with orthodox toolholders. The controls for operating the machine are grouped on the side so that the operator may reach all of them without undue movement.

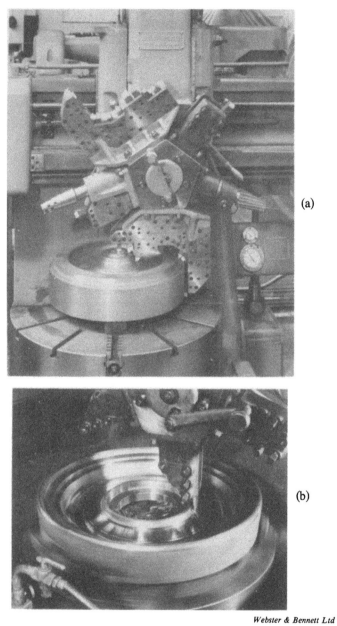

(a)

(b)

*Webster & Bennett Ltd*

Fig. 229 Examples of work on the vertical boring mill

For the type of work to which the machine is applied the horizontal turntable has advantages over the usual layout of a lathe with vertical chuck face. Working on the horizontal plane facilitates loading, setting, clamping and operating. The layout lends itself better to the application of more than one cutting tool at the same time, an important point when many surfaces have to be machined. By having a horizontal table no restriction is placed on the size to which it may be constructed, and the largest machine in the range of Messrs Webster & Bennett has a table 3·66 metres in diameter. (The sizes of vertical boring mills range from 900 mm upwards.)

In its application the vertical boring mill is used to a large extent for turning and boring operations on large diameter components, and for boring on jobs more conveniently handled this way than on the horizontal machine. It is opposed to that machine is so far that a stationary tool operates on revolving work, whilst on the other machine a revolving tool operates on stationary work. For some reason not easy to account for, turning and boring with stationary tool and revolving work is more satisfactory than when the work is held still and the cutting tool revolves.

Two examples of machining being conducted are shown at Fig. 229. The set-up at (a) involves multi-tooling with special toolholders to cut the profile on one side of the flywheel. Fig. 229(b) is an action photograph of the tool operating on the formed profile inside a housing. Here the tool is being guided by the electronic profile turning equipment in which the path of the tool is controlled by a stylus moving against a master template of the form to be produced.

# 9   Grinding

Of all the processes we are considering, grinding has probably made the most rapid and substantial advances during recent years, and its development has led to noteworthy, if not revolutionary, changes in our ideas of machine-tool production, and in the degree of accuracy and finish possible of attainment. Where tolerances of the order of 0·025 mm were once considered to represent a good performance, it is now possible to finish to within limits of less than one-quarter of this amount. The improvements in accuracy have been accompanied by a raising of the standard of surface finish, and have been brought about without any sacrifice in production time or increase in cost. The effect of these changes has been to raise the quality of our engineering products and through these the products of other industries, since most manufactures rely on various forms of machinery and engineering construction for help in their processes.

A simple example of the benefit gained by the ability to work to finer degrees of accuracy and finish is given by a journal bearing. Let us assume a 30 mm hole and shaft, the hole being finished to $30^{+0·03}_{+0·00}$ mm and the shaft to $30^{-0·02}_{-0·05}$ mm diameter for a running fit, and let the finish be such that the initial wearing period will result in an additional clearance of 0·02 mm taking place before the combination settles down. The smallest possible shaft in the largest possible hole will have a clearance of 0·08 mm, which will be increased to 0·10 mm when the initial wear has taken place. If, now the standard can be raised in such a way that the hole can be finished to $30^{+0·02}_{+0·00}$ mm, the shaft to $30^{-0·015}_{-0·04}$ mm, with an initial wear of 0·01 mm, the maximum possible working clearance, after initial wear, is reduced to 0·07 mm, an improvement of 30%. If conditions such as our second case can be instituted and maintained throughout the product, the result, according to the function of the article, may be increased accuracy, longer life, quieter operation, increased speed and/or efficiency, and so on.

It is difficult to say whether these improvements have come as a result of initiative on the part of the makers of grinding machines and grinding wheels, or whether their hand was forced by the users of their equipment. Be it as it may, the improvements have been made largely possible by developments in the technique of manufacturing the grinding wheel, and in the design of the grinding machine without any radical changes in the principles of the process itself.

Naturally, we are most interested in the applications of grinding in the workshop, and applied to the metals commonly used in engineering construction. We ought not to forget, however, that as a process it extends far beyond engineering in its scope. Grinding wheels are used for cutting, surfacing, shaping and other operations concerned in manufacture, throughout a wide range of industries. A selection of the materials and articles to which the process is applied is: bricks, carbon, concrete, cork, glass (cut, lenses, tumblers, tubing, mirrors, etc.), granite, marble, pearl buttons, porcelain, rubber, slate, pulping stones for newsprint, and so on.

### The grinding wheel

Grinding with modern abrasive wheels is just as much a metal-cutting process as any of the others with which we have dealt, except that the wheel has thousands of tiny cutting edges instead of the few large edges possessed by other rotary cutters. The old-time whetstone did its work by friction and abrasion, the residue from its use consisting of a mixture of metal dust with sand from the stone. If the grindings from a modern wheel working on a ductile material are examined they will be seen to be related in shape to short turnings from a tool (Fig. 230). This modern conception of grinding seems hard for some to adopt, and they still refer to the wheel as the 'stone', in the same way as the layman refers to a gear as a 'cog'.

A grinding wheel is made up of particles of a hard substance called the *abrasive* embedded in a matrix called the *bond*. The combination may be looked upon in a rough way as bricks and mortar; the abrasive grains constituting the bricks, and the bond holding them together as the mortar. The abrasive grains are second only to the diamond in hardness. Projecting from the surface of a wheel are thousands of edges of these grains which, when applied to the work, act in the same way as tiny cutting tools (Fig. 231). Naturally, as all tools eventually become dulled by use, the keen edges of those grains which are cutting will, in time, lose their cutting effect and at that stage, if the wheel is suitable for its work, they will either split and thus form a new edge, or break away from the wheel and expose another grain to carry on with the work. As we shall discuss later the essential

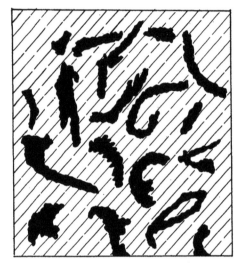

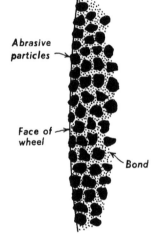

Fig. 230 Swarf from a modern grinding wheel, greatly magnified

Fig. 231 Showing edges of abrasive grains projecting from face of wheel

of grinding wheel selection is to ensure that the grains of abrasive do not leave the wheel before they are dulled, or stay in it too long after their keen edges have been taken away. The first fault causes wheel wastage and the second gives rise to a defect known as *glazing*.

## The abrasive

Until the discovery of methods of making artificial abrasives about 100 years ago the two minerals used as the cutting agent in grinding wheels were *emery* and *corundum*. Both of these are impure forms of aluminous oxide, emery consisting of crystals of the oxide embedded in a matrix of iron oxide and corundum consisting of aluminous oxide associated with varying amounts of other impurities. Emery is about 60%, and corundum 90% pure. Modern grinding wheels are made with the artificial abrasives we are about to describe, and although they are still often called emery wheels it is doubtful whether any such now exist except in museums. Corundum is still used as an abrasive in some industries. An important characteristic of the artificially made abrasives is their purity, this having an important bearing on their efficiency. These abrasives are:

(1) *Silicon carbide* and (2) *aluminous oxide*.

Silicon carbide is a chemical combination of carbon and silicon and is made in a large rectangular resistance type electric furnace. In making the abrasive the furnace is charged with coke, sand, salt and sawdust. The coke supplies the necessary carbon and the silicon is derived from the sand. The sawdust is added to render the mass porous so that the gases generated may escape. Salt is added to eliminate any impurities such as iron, which it does by volatilising, taking them as chlorides. In the furnace the action takes about 36 hours at approximately 2000 °C, and when the furnace has cooled down the silicon carbide is taken out in the form of large masses of crystals. These are broken up, crushed and rolled into small grains which are screened and graded, cleaned, washed and dried. Silicon carbide is greenish black in colour.

Aluminous oxide (fused alumina) is also prepared in an electric furnace from bauxite (hydrated aluminium oxide), a clayey mineral which is found in many parts of the world. For the process a cylindrical arc furnace is used consisting of an outer shell resting on a base and provided with two electrodes for supplying the current. In operation a layer of bauxite is introduced and the electrodes lowered until they rest on it. A path of graphite is then laid between the electrodes to serve as the initial passage for the current, but as soon as the charge is molten it forms a good conductor in itself. When the first layer of bauxite is molten another layer is introduced and the electrodes raised. This procedure is repeated until the furnace is full, the whole process occupying about 36 hours at a temperature of about 1800 °C. When the action is finished and the furnace cooled sufficiently the outer shell is removed to allow the ingots of alumina to be withdrawn. These are then broken up and treated as before. Aluminous oxide is a reddish-brown colour.

Both abrasives are graded by sieving through screens having holes or meshes graded in size. The *grit* signifies the number of meshes used to grade any particular size. The screens from 200 to about 46 are generally made of silk, whilst those from 46 to 8 are of wire. Finer than 200 are the 'flour' abrasives designated as F, FF, and FFF. These and finer powders used by jewellers, etc., are separated by a water flotation process and in some cases graded under microscopic control. In comparing the two abrasives (both, of course, are able to cut the hardest steel), silicon carbide is harder and more brittle than alumina, for this reason it is more suitable for operating on materials of low grinding resistance (low in hardness and toughness) which dull, but do not readily fracture alumina. As silicon carbide is more easily fractured new cutting points are provided without undue wheel wear when grinding low resistance materials. Alumina abrasive is the more suitable for most steels because of its greater toughness to cope with the increased grinding resistance offered. Naturally, there are materials for which either abrasive is suited, cast iron being one metal in this category (see Fig. 235). The different makers of grinding wheels have their own trade names for carbide and alumina wheels, but the nature of the name gives the clue at once to the abrasive, e.g. Carborundum, Carbolite, Crystolon, etc., are silicon carbide wheels, whilst Aloxite, Alundum, Oxaluma, etc., belong to the aluminous oxide group.

### The bond

The bond is the substance which when mixed with the abrasive grains holds them together enabling the mixture to be shaped to the form of the wheel, and after suitable treatment to take on the necessary mechanical strength for its work. The degree of hardness possessed by the bond is called the *grade* of the wheel, and indicates the ability of the bond to hold the abrasive grains in the wheel. A soft bond permits the grains to break away more readily than a hard one, and should be used where the abrasive becomes rapidly dulled, i.e. when grinding hard materials. A hard bond retains the abrasive grains longer and should be used on soft materials— hard wheel for soft work and soft wheel for hard work. It should be remembered that the term hardness or grade of a wheel has no relation to the abrasive material used in its construction; a soft graded wheel can be made of the hardest of abrasives. There are several types of bonding materials used for making wheels, the chief being: (1) Vitrified, (2) Silicate, (3) Shellac, (4) Rubber, (5) Synthetic Resin, (plastics).

**Vitrified bonds.** The vitrified process is used for the majority of wheels and is responsible for probably $80\%$ of the output. Wheels made by this method can be supplied in a wider range of grades than in any other, and are used on more varied classes of grinding work. The bonding materials used consist of various kinds of fusible and refractory pottery clays with which the grains are thoroughly mixed. The forming of the wheel may be carried out either by the Puddled or the Pressed Process. In the puddled process the correct proportion of grain and bonding material are mixed wet and poured into a mould to dry. The wheel is then shaped on a machine operating on the principle of a potter's wheel. After this the wheels are charged into a kiln for the burning process which occupies 2–3 weeks; one week to attain the correct heat, a suitable period at the heat and about one week to cool down. In the pressed process the grains and bonding clay are mixed in a semi-dry state and the wheel moulded under pressure. By this method wheels can be made under better control as regards density, giving a wider range of grades, and the process, in time, will probably supersede the puddled process.

**Silicate bonds.** Silicate of soda (water glass) is the chief bonding agent used for these wheels, oxide of zinc being added as a waterproofing agent. The abrasive and bonding materials are first mixed thoroughly in mechanical mixers, and then the mixture is tamped in iron moulds to shape the wheel. The shaped wheel is then baked at a low temperature for 2–4 days to set the bond. Silicate wheels have a milder action and cut with less harshness than vitrified wheels. For this reason they are suitable for grinding fine edge tools, cutlery, etc. When required to order they can be made in a much shorter time than vitrified wheels and, owing to the lower baking temperature (260 °C), can be moulded in or on the special steel adaptors designed to fit certain machines. They lack the mechanical strength of vitrified wheels, however, and are easily damaged.

**Shellac bonds.** Shellac wheels may be made to 3 mm or less in thickness. Flake or powdered shellac is mixed with the abrasive and heated until the melted shellac has coated each grain. The mixture is then rolled into sheets of the required thickness and the wheels cut out with a die. Thick wheels are tamped in steel moulds and then pressed. The final process is a mild baking with the wheels packed in sand.

**Rubber bonds.** Rubber wheels are made by taking a sheet of pure rubber, to which sulphur has been added as a vulcanising agent, spreading the abrasive on it, folding the sheet upon itself and passing it through rolls. More abrasive is then added, the sheet again folded and rolled. This process is repeated until the desired quantity of abrasive has been thoroughly kneaded into the rubber, when the wheels are cut out of the sheet and vulcanised under heat and pressure.

**Synthetic resin bonds.** Wheels of this type are made with one of the resin (plastics) materials as the bond. Before cooking, the resin is in the form of a powder which is thoroughly mixed with the abrasive grains and pressed into moulds the shape of the wheel. Whilst the pressure is applied the mould is heated and this brings about a chemical change, hardening the bond to form the type of material of which plastics articles are made.

Elastic wheels, because of their close, dense bond which nullifies free cutting, cannot be classed as 'production' wheels. The amount of material they will grind away in a given time is small and if they are forced the heat generated will soon destroy them. Their chief field of application is when an extremely thin or fine wheel is required for such purposes as cutting off, slitting, finishing fine work such as form tools, and where an extremely fine finish is required as on cams, rolls, ball races, etc. Plastics bonded wheels are used extensively for fettling the rough spots off steel and malleable castings. Because of the higher mechanical strength of their bond elastic wheels may be operated at a higher speed than vitrified wheels and their best efficiency is obtained at approximately twice the speed that would be employed for vitrified or silicate wheels.

### Grit and grade

According to the application of a particular wheel, its grit, grade and abrasive must be varied. The grit, as we have seen, is the size of the abrasive grain and is specified according to the meshes of the grading sieve. The screened sizes generally made vary from about 8 to 240. Following these are the unclassified flours F, 2F, 3F, 4F and XF, and then the classified flours ranging from approximately 300 to 900.

The grade, referring to the hardness of the bond, is usually designated by the letters of the alphabet. By this classification the bond hardness is graded from A (very soft) to Z (very hard) although the main range of wheels is covered by the letter series, E (soft) to U (hard). Under this classification a 46K wheel would have a 46 grit in a soft-medium bond whilst an 80R would have a grit of 80 in a hard bond. Until quite recently different wheel makers had their own methods of grading but The Abrasive Industries Association have now adopted a standardised system which is observed by the leading makers of wheels (see later).

### Wheel structure

The proportion of bond in a wheel varies from about 10% to 30% of its total volume, and the structure of the wheel will depend upon this proportion. If it is high the spacing of the grains will be wide as shown at Fig. 232(a), whilst a low proportion of bond to abrasive will give a close structure as shown at (b). The structure, even with identical bonds and abrasives, will influence the grinding action, open structures tending to have effects similar to soft bonds and vice versa. Wheel manufacturers are now controlling the structure of their wheels and adopting a system of classification. When structure is classified the numbers 1 to 15 cover the range commonly in use, low figures referring to close, and high to open grain spacing.

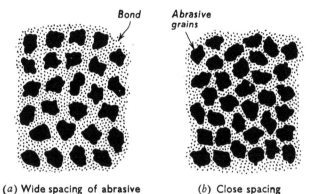

(a) Wide spacing of abrasive      (b) Close spacing

Fig. 232 Showing variation in structure possible with similar grit and bond

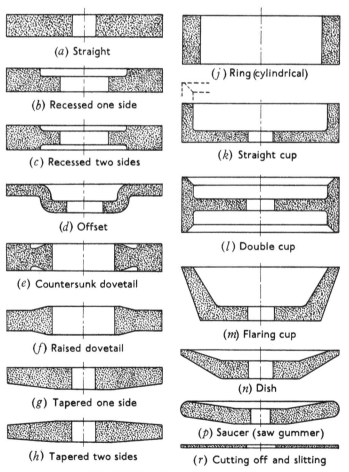

(a) Straight

(b) Recessed one side

(c) Recessed two sides

(d) Offset

(e) Countersunk dovetail

(f) Raised dovetail

(g) Tapered one side

(h) Tapered two sides

(j) Ring (cylindrical)

(k) Straight cup

(l) Double cup

(m) Flaring cup

(n) Dish

(p) Saucer (saw gummer)

(r) Cutting off and slitting

Fig. 233 Grinding wheel shapes

## Wheel shapes

Grinding wheels are made in a wide variety of shapes to meet an immense range of work. In addition, various makes of machines have special wheel adaptations which add further to the problem of wheel makers. It would not be possible in the space available for us to discuss all the shapes that are made, but at Fig. 233 are shown a selection of the most common. The shapes (a) to (h) are used as disc wheels (grind on periphery) on cylindrical, surface and general purpose tool, etc., grinding machines. Wheels (j) to (l) are mostly used on cup wheel surface grinders. Shapes (m), (n) and (p) are used for tool and cutter grinding, whilst the thin wheel (r) is for slitting and cutting off. Often on the larger cup wheel surface grinding machines instead of a solid wheel a number of segments are fitted to a chuck, and an example is shown at Fig. 234. This form of construction considerably reduces wheel cost as the set of segments is much cheaper than a solid wheel. The gaps between the segments also promote freer cutting.

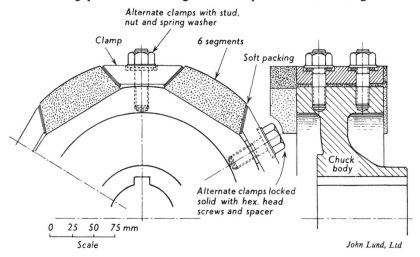

Fig. 234 Segmental chuck

### Wheel selection

The problem of selecting a suitable abrasive, grit and bond for the various types of work likely to be encountered is complex in view of the wide choice of wheels offered, and the range of materials used in manufacture to which the grinding process is applied. We will content ourselves with discussing the fundamental principles concerned, reminding our readers that the grinding wheel manufacturers have an efficient staff of experts whose job it is to study the whole aspect of grinding, and who are always ready to place their experience at the disposal of a customer. When ordering a wheel and the most suitable grit and grade is in doubt, rather than guess, or assume knowledge that is not there, it is better to advise the maker for what machine, purpose and material the wheel is required, leaving him to supply the most suitable type. On a general purpose machine the use of the most suitable wheel is cramped by the impossibility of making continual changes of 'wheel as changes occur in the type of material being ground. To overcome this as far as possible makers can supply two or three general purpose wheels to give a good all-round performance on a specified range of materials (e.g. one for all classes of steelwork and another for cast metals). On many machines wheel changing is facilitated by detachable collets which hold the wheel permanently, and the collet complete with wheel is taken off, to be replaced by another collet complete with wheel. This saves time and the wheel is not disturbed on its seating. In addition, a wheel which was true when removed will be the same when put on again.

**The abrasive.** As we have already explained, although either of the abrasives is hard enough to grind most of the materials with which we have to deal, the characteristics of silicon carbide render it the more suited to grinding materials of low tensile strength, whilst alumina is most efficient when applied to hard, tough materials. This is illustrated by Fig. 235, which also shows, in the neighbourhood of where the lines cross, the materials for which either abrasive is suitable. The line for emery has been put on and indicates that it never gives more than medium efficiency, which is accounted for by the dilution it suffers on account of its impureness.

The grit size will largely influence the finish given to the work and the freeness with which the wheel cuts. A large grit will promote an open, free-cutting wheel-face and enable stock to be removed more rapidly, but will not leave such a fine finish as a smaller size of abrasive grain.

**The bond (grade).** The determination of the most suitable grade of wheel for any particular job is not nearly as straightforward as that of choosing the abrasive since so many factors influence it. The first and general rule is '*a hard wheel for soft work and a soft wheel for hard work*'. Grinding hard material dulls the grains more rapidly than soft work, so that the softer bond is necessary to permit the grains more easily to split up or leave the

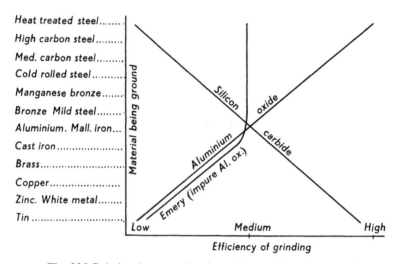

Fig. 235 Relation between abrasive and material to be ground

wheel for the purpose of exposing fresh grains to carry on. Soft material has a less rapid dulling effect and a harder bond is permissible to hold the grains longer. Additional factors which influence bond selection are as follows:

1. *The arc or area of contact.* The two extremes of arc of contact are shown at Fig. 236(a) and (b) for a disc wheel, and of area of contact at (c) and (d) for a cup wheel. For large arcs or areas of contact the wheel must be softer than when the zone of action is small. (Note that the *width* of the wheel-face has no effect on this provided the pressure per unit area remains the same. There should be no difference between the actions of disc wheels 12 mm wide and 50 mm wide, or between cup wheels with similar variations in rim width.)

2. *Work speed.* A high work speed means more material ground in a given time and greater wear on the wheel. Hence the higher the work speed, or the harder the wheel is forced in offhand grinding, the harder must be the bond.

(Offhand grinding is when the work is held in the hand as for rough, approximate work and often for sharpening cutting tools.)

3. *Type and condition of machine.* Light machines subject to vibration and machines in poor condition require harder wheels than heavy, rigidly constructed and well maintained machines. Loose spindle bearings, shaky foundations, etc., cause much trouble in grinding because hard wheels have

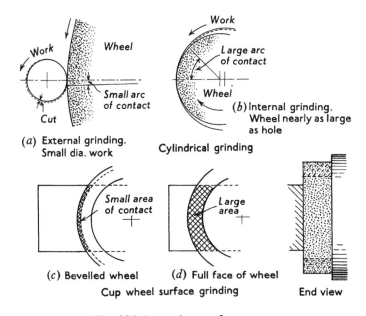

Fig. 236 Arc and area of contact

to be used to overcome rapid wheel wear and thus cutting efficiency is sacrificed.

4. *The personal factor.* A skilled operator can work with softer wheels than one who is not so skilled and hence obtain better and more economical production. Piece-work grinding usually necessitates the use of harder wheels than when day work is the rule. It has been found that on offhand grinding, costs may vary as much as 100% on the same work, with the same type of machine in the same factory.

5. *Use of a coolant.* If grinding takes place without a coolant a softer and freer cutting wheel will be necessary than if cooling water is used. Hence the use of a coolant enables a harder wheel to be employed.

**The process**

The selection of the wheel-making process is influenced by the following considerations:

1. Very thin wheels or those subjected to bending forces when in use should be made by the shellac, bakelite or rubber process.

2. For very high finish where rapid cutting is not important use shellac or rubber wheels.

3. Use silicate wheels to replace sandstones on cutlery, etc., and when the wheel is very large, or required quickly to special order.

4. For general purposes, and maximum cutting efficiency, use vitrified wheels.

## Wheel classification

The Abrasive Industries Association have adopted a British Standardised Marking System for grinding wheels. This will be of great benefit as previously the position was rather chaotic with different makers using their own symbols. The system, illustrated below, specifies the abrasive, the grain, the grade and the bond. The example given represents an aluminium oxide abrasive of 46 grit in a vitrified bond of medium hardness and structure. The Prefix, the Structure and the Suffix (shown in smaller print in the diagram) are optional in their use. The structure we have already discussed and the other two headings are for specifying individual manufacturer's abrasive and wheel types (e.g. W stands for white abrasive, etc.).

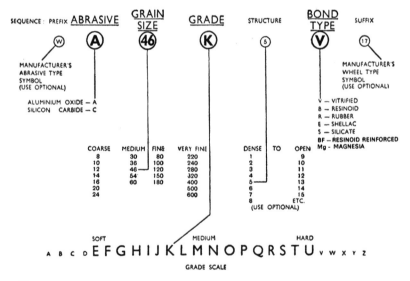

The tag or label attached to a wheel should be preserved as the information on it will be of value when ordering a replacement.

## Wheel mounting

In the interests of satisfactory operation and safety it is important that grinding wheels should be mounted correctly on the machine, and before mounting examined for any defects. The wheel should first be examined for any flaw or crack which, under the stress set up due to the high speed of revolution, might lead to the fracture of the wheel. Sound vitrified and silicate wheels have a clear, bell-like ring when struck, so that after a visual examination the wheel should be held with the fingers in the bore and struck lightly with the knuckles or a piece of wood (not steel). If any crack is present the fact should be indicated by the sound.

**Wheel balance**

The second point which deserves attention on all grinding operations except the very roughest is the balance of the wheel, and before mounting the wheel should, if possible, be tested for balance. This can be carried out with the wheel mounted at the centre of a perfectly straight and round spindle, the assembly then being rested on level knife-edge ways on a lathe bed or on a special stand (Fig. 237). For the test to be really satisfactory the wheel should be mounted on its own spindle and the test made after truing up the wheel-face. Any out of balance will result in the wheel coming to rest with the heavy side underneath, and the error may be rectified by cutting some of the lead from the bush with which the bore of the wheel is provided, lead being removed from the heavy side. If the amount out of balance is too great to be corrected in this way it may be necessary to dig away some

*The Churchill Machine Tool Co, Ltd*

Fig. 237 Grinding wheel balancing stand (static)

of the wheel itself, but this should only be undertaken by an expert. At the speed with which wheels revolve, even small unbalanced amounts lead to considerable forces; for example, at 1500 metres per minute (the usual peripheral speed) a wheel which is out of balance to an amount equivalent to 10 grams at 150 mm radius will be subjected to an unbalanced centrifugal force of approximately 42 newtons (4·3 kgf). More often than not wheels are in correct balance when they are new, but as the wheel is reduced in diameter by wear it may lose its balance. Lack of balance, if serious, will involve the risk of the wheel bursting due to the stresses set up, but on precision grinding operations small out-of-balance effects may be so detrimental to the accuracy of the surface and the finish produced as to spoil the work. For high-class precision machines a *balancing collet* may be used which fits the spindle and holds the wheel in the usual way, but is provided with adjustable segments which may be moved about for the purpose of obtaining true balance.

Grinding wheels are held to their machine spindles by fitting with the bore of their lead liner bush and being clamped between two flanges. This is shown by Fig. 238, and the following points are important:

1. The sides of the wheel and the flanges which clamp them should be flat and bear evenly all round.

2. The lead bushing should be an easy fit and no force used to get the wheel on. (Scrape out with a knife if necessary.)

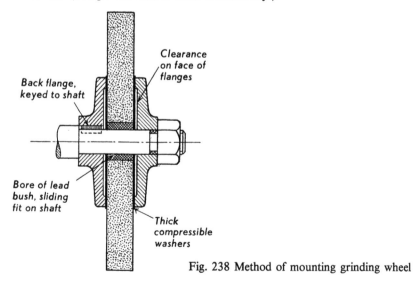

Fig. 238 Method of mounting grinding wheel

3. The back, fixed flange should be keyed, shrunk or otherwise fixed to the spindle. If it is not, there is no definite method of transmitting power from the spindle to the wheel.

4. Both flanges must be relieved so that they only bear on the wheel at their rim. On no account must one or both of them touch the sides of the wheel anywhere else. Blotting paper, rubber washers or some compressible material should be used between the flanges and the wheel.

5. Both flanges should be the same diameter, at least one-third and preferably one-half the wheel diameter. If they are not equal in size bending stresses will be induced in the wheel.

6. The nut should be tightened only enough to hold the wheel. Undue tightness is unnecessary and undesirable. As soon as a wheel has been fitted, and before the machine is started up, the wheel guard should be replaced and screwed up. After fitting a new wheel start up the machine and stand clear for about a minute. Under the Abrasive Wheels Regulations 1970, which came into force in April 1971, there are several requirements not included in previous Factory Acts. One of these is that persons who mount certain types of wheel must be trained and certified. Particulars of these Regulations may be obtained from H.M. Factory Inspectorate.

**Wheel truing and dressing**

As soon as a fresh wheel has been fitted it will be necessary to true its face and probably its sides for a short distance down. Truing or dressing is also necessary from time to time during the course of working to correct for uneven wear and to open up the face of the wheel so as to obtain efficient cutting conditions. The two terms have by no means the same significance as, whilst truing usually implies rendering the wheel face (or sides) perfectly true, dressing is a process of cleaning and opening up the face although not necessarily rendering it true.

Truing is carried out with a diamond which has a shearing action on the abrasive grains and the bond, and so removes dulled or irregular groups of grains. It is able to do this because it is harder than the abrasive, although not much harder than silicon carbide which wears it more than alumina. The diamonds most commonly used are South African Brown Bort stones. The Grey Bort and the Ballas, although superior in hardness, are more expensive and only used on large wheels where their increased hardness gives longer life against the additional wear. To render it suitable for its work a diamond must be mounted in a holder. This is done by peening, brazing or securing the diamond by casting a low melting point metal round it, leaving sufficient of the stone protruding for it to act as a cutting edge. Recommended sizes of diamonds are as follows: up to 150 mm wheel, $\frac{3}{4}$ carat; 150 mm to 300 mm wheel, 1 carat; 300 mm to 450 mm wheel, $1\frac{1}{2}$ carat; 450 mm to 600 mm wheel, 3 carat. Except on saucer and irregular-shaped wheels used for cutter grinding the diamond holder should be clamped to the machine (Fig. 239) and brought slowly up to the wheel-face. (Nothing should ever be allowed to bump against a wheel.) The rate at which the stone is taken across the wheel as well as its sharpness or otherwise affects the wheel condition. For roughing, a relatively rough open wheel-face is required and to produce this the diamond should be moved relatively fast and should, if possible, have a fairly sharp point. When a fine finish is required (at the expense of efficient cutting) a blunt diamond traversed slowly across the wheel-face will give suitable conditions but too slow a movement may glaze the wheel.

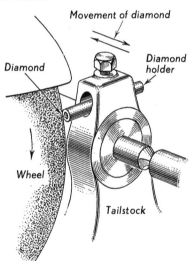

Fig. 239 Truing wheel of cylindrical grinding machine with diamond clamped to tailstock

Actually, if a wheel of the correct grade is being used a blunt diamond is best, and if the diamond has a very sharp point it may not be possible to render the wheel-face smooth enough to give a good finish. As soon as the diamond is heard to touch the wheel it should be traversed across by power-feed of the table and then advanced about 0·02 mm each time until the sound indicates that the wheel-face is round and flat. When the grinding is wet the wheel should be trimmed wet and a good flow of water minimises wear of the diamond as well as preventing its cracking. On dry machines use the diamond dry, but take care against undue heating of the stone and the risk of its fracture or the loosening of the mounting. Readers who have experienced the loss of a diamond and have been compelled to search until they have found it will know how much more easily they are lost than found again.

For hand grinding and certain classes of machine grinding in which a precision finish is not necessary, the exact trueness imparted by the diamond is superfluous, and to use it for such purposes causes unnecessary wastage of an expensive tool. In such cases wheels may be dressed with one of the various tools made for the purpose. Many of these have a number of star-shaped wheels which revolve when pressed into the wheel-face and

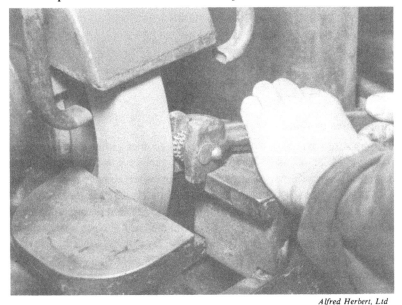

*Alfred Herbert, Ltd*

Fig. 240 Dressing the side of a wheel with a star wheel dresser

by so doing dig away the bond and release the dulled grains (Fig. 240). Sometimes wheels are dressed with a solid lump of abrasive crystals, or a small wheel of the same material mounted on a spindle with handles at each end for holding.

**Wheel speeds**

The grits and grades recommended by wheelmakers for specific operations pre-suppose that wheels are operated within certain prescribed speed limits. If this is not possible then the grade of the wheel must be adjusted to suit the revised conditions.

AVERAGE RECOMMENDED WHEEL SPEEDS IN SURFACE METRES PER MINUTE

| Cylindrical grinding | 1675–2000 | Wet tool grinding | 1520–1820 |
|---|---|---|---|
| Internal „ | 600–1200 | Cutting off with rubber, | |
| Surface „ | 1200–1520 | shellac and bakelite wheels | 2750–3650 |

To find the rev/min for a grinding wheel the same expression may be used as for cutting speed, for example, a 250 mm wheel to run at 1820 surface metres per minute:

$$N = \frac{1000S}{\pi d}$$

where $N$ = rev/min, $S$ = surface speed (m/min), $d$ = wheel diameter (mm)

$$N = \frac{1000 \times 1820}{3 \cdot 14 \times 250} = 2320 \text{ rev/min}$$

**Maintenance of speed**

It is important that as a wheel wears and becomes reduced in diameter its speed should be adjusted to maintain the surface speed at approximately the correct amount. For example, a 250 mm wheel to have a surface speed of 1820 m per minute should run at approximately 2320 rev/min. Normally this wheel would be used down to 200 mm or 225 mm diameter, and if still running at 2320 rev/min when 225 mm diameter its surface speed will only be 1640 m per minute. If it is to operate at its original efficiency its speed should be raised to about 2580 rev/min. This point is often overlooked, particularly on general purpose shop grinders which one sees running at the same speed irrespective of the size to which the wheel has been reduced.

**Operating faults—loading and glazing**

There are two common faults which occur with grinding wheels. Loading is when the interspaces between the abrasive grains become clogged with particles of the material being ground, with the effect that eventually the cutting edges do not project sufficiently to do their work. Loading may be caused by forcing the cut too fast or too deep to allow the chips to be carried away or by using the wheel to grind materials softer than for which it is suited. Fine grained and soft bonded wheels do not load as readily as their opposites and an increase in speed may help to effect a cure.

Glazing occurs when grains which have lost their sharpness are still retained by the bond. It is caused either by using a wheel which is too hard, by too high a wheel speed, or too low a work speed. It may be reduced by increasing the work speed and reducing the cut to promote more rapid disintegration of the wheel. When a wheel becomes loaded or glazed it should be trimmed or dressed immediately, and if the fault persists speed and wheel suitability should be checked up with a view to effecting a cure.

A fault the opposite to loading and glazing is that of excessive wheel wastage. This may be reduced by reducing the work speed and increasing the cut.

*Alfred Herbert, Ltd*

Fig. 241 Floor stand grinder

## General hand grinding—floor stand machine (Fig. 241)

Almost every shop is equipped with one or more stand grinders for general use such as tool grinding and such other odd jobs as may be necessary. For the general hand grinding of tools a vitrified alumina wheel of about 30 grit in a medium bond is most suitable. These machines are fitted with rests for supporting the work whilst it is being ground, and these should be adjusted so as to be as close to the wheel as possible to avoid the danger of the work slipping down and becoming wedged between the rest and the wheel. On some of these machines the design provides a stream of water, sometimes actuated by raising a tank underneath so that the wheel dips into it. Actually this is not very satisfactory, as even if the contraption is in working order the stream of water is totally inadequate and is, in fact, more detrimental than grinding dry, because if the flow of water is insufficient to keep the tool from heating up, a stream of water on the hot tool causes surface cracks. It can be assumed, therefore, in general, that grinding on these machines is dry, and to avoid burning the tool it should not be pressed too hard, and should be manipulated with a rocking motion from side to side across the entire face of the wheel. The tool should always be

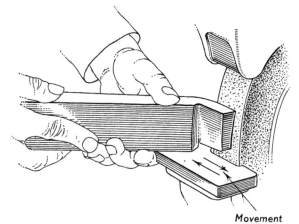

Movement

*The Norton Grinding Wheel Co, Ltd*

Fig. 242 Hand grinding a turning tool

moved about, for the additional reason of equalising wheel wear (Fig. 242). Overheating is further minimised by keeping the wheel-face well dressed. When grinding tools this way by hand, exact machine-like precision in the rake and clearance angles cannot be expected, but with the help of a few simple angle gauges satisfactory results are possible. A good plan during early attempts is to obtain some tools ground up by an experienced operator and to use these as models from which to copy until facility is gained in holding and manipulating the tool. The tool should be ground so that the wheel always travels towards the cutting edge, and if it is to be used for fine finishing or cutting soft metals its edge should be oilstoned lightly after grinding, as the finish is influenced by the keenness of the cutting edge.

Many tools are in use which consist of bits held in a separate shank. These are often held at a sufficiently large angle to give as much top rake

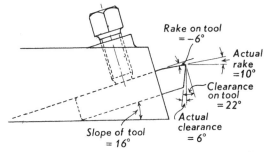

Fig. 243 Effect on cutting angles of initial slope of tool

as is ever required, and when grinding the bit allowance must be made for the angle at which it is set. A larger clearance will be required and some of the rake may have to be taken off by putting some negative rake on the tool. For example, if the bit is inclined at 16° in the holder and a tool point is required with 10° top rake and 6° clearance the point when ground must have $10 - 16 = -6°$ top rake and $16 + 6 = 22°$ clearance (Fig. 243).

## Chip breaking

Long, curly chips from a lathe tool are inconvenient as well as dangerous, and it is possible to grind the tool so that the chips will be broken up into

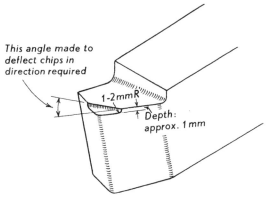

Fig. 244 Chip-breaking groove on turning tool

short curly lengths as they pass over the tool face. Long continuous chips are more common when turning soft and tough steels; no trouble is experienced with brittle steels and cast iron. One method of preventing the formation of these chips is to grind a small groove across the top face of the tool to about 3 mm to 6 mm back from the cutting edge. The groove need not be very deep and it has the effect of breaking up the chip into short lengths. On large production turning machines special chip breakers are incorporated in the design of the cutting tools, as without them the mass of stringy turnings would become unmanageable and would become entangled with the machine, preventing its operation (Fig. 244).

### Special tool grinders

When large numbers of tools have to be ground and uniformity is important hand grinding becomes out of the question. For such purposes there are several makes of tool grinders which incorporate facilities for holding and locating the tool shank, and setting it to produce the required cutting angles on the nose. One well-known example of such machines is the 'Lumsden' grinder.

### Grinding on the lathe

Some form of grinding attachment is necessary for use on the lathe if only to finish off the points of the centres after they have been hardened.

Some attachments take the form of a small electric motor provided with a shank which is clamped to the tool-post, the grinding wheel being carried on an extension of the motor armature spindle. Because of their method of attachment such machines are called 'tool-post grinders'. On a more elaborate pattern the motor is carried at the front of a baseplate and drives the spindle by means of an endless woven belt. The spindle is carried at the other end of the baseplate and the whole assembly bolts to the top of the compound slide (Fig. 245). On the better types provision is made for adapting an internal grinding spindle so that holes may be ground.

When using the tool-post grinder for finishing centres the compound slide should be set at 30° with the axis of the spindle and the attachment set up with the motor spindle approximately at the same angle and at the same horizontal level as the lathe centres. The centres, in turn, should be carefully cleaned on their taper shanks and inserted in the spindle for grinding. The lathe should be run at a medium speed, the grinding cut traversed by hand with the compound slide and extra cut put on with the cross-slide. It is advisable, with the centre which is for permanent use in the

*Wolf Electric Tools, Ltd*

Fig. 245 Lathe grinding attachment operating on a component

headstock, to mark a line on it which will serve as a guide for always assembling the centre in the same peripheral position as that in which it was ground true.

Lathe grinding attachments have their uses and are occasionally the means of enabling a job to be done which might otherwise present difficulties. For accurate grinding, however, they are of no use since they possess none of the qualities of design, rigidity, accuracy, etc., necessary for any but makeshift jobs. Their makers never intended them to be other than a useful accessory for the lathe, and it should never be considered possible to supplant the cylindrical grinding machine even with the best of such attachments.

## Cylindrical grinding

The work of the cylindrical grinder consists of finishing the faces and diameters of cylindrical work previously turned in the lathe. Generally, the operations of external finishing and boring, which are both performed on the lathe, are treated separately for grinding, external work being done on the *cylindrical* grinder and bores on the *internal* machine. The *universal* grinder may be adapted to perform both external and internal work, but even so most bores are ground on an internal machine, as by designing such a unit for this sole purpose it may be provided with various features which promote increased efficiency. No doubt the process of cylindrical grinding was first developed to finish the diameters of work that had been hardened, and was therefore beyond the cutting capacity of the lathe. There are many today who still regard the finishing of hardened work as the function of the grinding machine, but modern developments in machine design and wheel manufacture have given the process a wider application than this, and for work within its scope it may now be regarded a superior and cheaper method of finishing, whether the metal be hard or soft. As compared with turning, the facility with which an accurate size may be obtained, and the inherent accuracy of the process itself, have proved that it is often better, and cheaper, to turn a job to within $\frac{1}{2}$ mm or so of size, and then take it to a grinding machine for finishing.

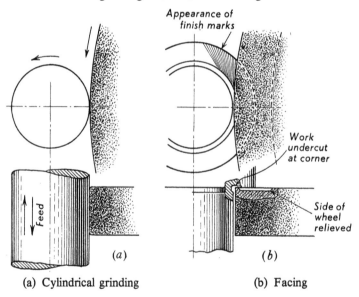

(a) Cylindrical grinding          (b) Facing

Fig. 246

**External work**

External grinding is performed with the periphery of a disc wheel as shown by Fig. 246(a). The facing of ends and shoulders may be carried out with the side of the wheel as at (b), after relieving the wheel so that only a narrow ring at its outside does the cutting. When the work is carried on centres, and this is more the rule than otherwise, the arrangement differs

SPEED AND FEED OF WORK

from the lathe insomuch that it rotates on two dead centres. Driving takes place through a dog somewhat similar to a lathe, except that the driving pin is carried on a flange which rotates on its own bearing, independent of the centre. The tailstock supporting the other end of the work is spring-loaded to compensate for temperature length changes. A diagram showing

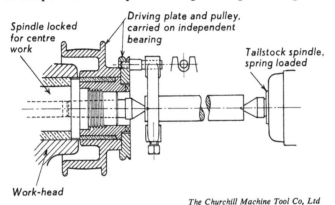

*The Churchill Machine Tool Co, Ltd*

Fig. 247 Supporting and driving centre work on grinding machine

the centres and method of driving is shown at Fig. 247. By causing the work to rotate with its own centre holes bearing on two dead centres, the ground diameters will be concentric with the centres and hence concentric with each other. On the lathe, where the headstock centre rotates with the work, the concentricity at that end is dependent on the trueness of that centre.

**Speed and feed of work**

The work speed in cylindrical grinding is influenced by the following factors: (1) arc of contact between wheel and work (depends on work diameter); (2) material hard or soft; (3) rate of feed and finish desired; (4) dimensions of work as it affects its tendency to heat up (thin work heats up rapidly); (5) tendency of work or machine to vibration.

On large diameters of work the greater arc of contact prevents the abrasive edges from penetrating to the same depth as when the work is smaller and the arc of contact less. For this reason the surface speed of large work should tend to be greater than for small. If the work is soft, or the wheel gives the impression of being hard, a higher work speed may be used. Slower speeds should be used for finishing, and to prevent the wheel-face from breaking down if a heavy feed is being used. Small diameter and thin hollow work needs to run fast enough to prevent any tendency for local heating up to take place, and this is minimised by keeping work speed and table travel as high as possible with plenty of cooling water. Vibration of work not only affects the finish obtained, but also tends to cause the wheel-face to disintegrate. It must be avoided at all costs by regulating the speed and using steadies. The tendency is most common on small diameter and hollow work. Glazing, which indicates a hard wheel, necessitates a higher work speed, whilst rapid wheel wear, indicating a soft wheel, calls for a reduction in speed. It has been found from experience that an average surface speed of 25 metres per minute is a suitable basis upon which to work, modifications being made to cope with special working conditions.

**Longitudinal feed.** The longitudinal feed of the wheel over the work must be considered in relation to the wheel width. The wheel does not feed slowly in the same way as a lathe tool, but for each revolution of the work moves a considerable fraction of its width. This should be approximately one-half to two-thirds the wheel width and the effect on the face wear of the wheel is shown at Fig. 248. At (a) the feed is $\frac{2}{3}$ the wheel and, as will be seen, work movement to the right involves the portion AC of the wheel cutting,

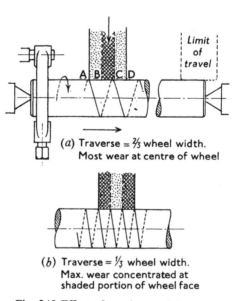

(a) Traverse = $\frac{2}{3}$ wheel width.
Most wear at centre of wheel

(b) Traverse = $\frac{1}{3}$ wheel width.
Max. wear concentrated at shaded portion of wheel face

Fig. 248 Effect of varying work traverse

whilst movement to the left brings the portion BD of the wheel-face into action. The centre third BC of the wheel, therefore, does twice as much cutting as the outer portions and the wheel tends to wear concave. The conditions shown at (b) are with a traverse of $\frac{1}{3}$ the width, and in this case the outer thirds only do most of the work, causing the face to wear convex in shape. The wheel must not be allowed to run off at the ends of the work, but the reverse should take place when it is about one-third off as shown dotted.

**Radial (in) feed.** Reduction in the work diameter is accomplished by an in-feed of the wheel which should take place at each end of its travel. This feed should vary between about 0·005 mm to 0·04 mm according to the size of machine, type and rigidity of work, etc., and on most machines facilities are incorporated for its automatic application at each end of the table travel. Generally, some form of pawl motion is used, operating on the wheel-head actuating shaft, with provision for tripping the motion when the desired size has been reached. The stopping point of the feed should be controlled so that the work is brought to its final size by several passes of the wheel after the last increment of feed has been applied. When working to very fine limits it is not advisable to trust the feed trip but to apply the final increments of feed by hand.

### Plunge cut grinding

When the wheel width exceeds the length of work to be ground, the wheel, after being trimmed up parallel, may be fed straight in with no longitudinal feed of the work as shown at Fig. 249. It is an advantage, for the sake of equalisation of wheel wear, to rock the work slightly (about 3 mm) longitudinally whilst the cut is taking place.

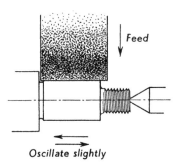

Feed

Fig. 249 Plunge cut grinding        *Oscillate slightly*

## Facing

When the side of the wheel is used for facing ends and shoulders the face should be brought up to the wheel by hand and then the machine table carefully inched along until the wheel touches the work. After this, cut is applied gently by tapping the traverse wheel gently with the hand. Care and light manipulation is necessary for this operation, as when the wheel has been relieved to allow a thin band to project at its edge this is very tender, and if brought heavily against a shoulder will be destroyed (Fig. 246(b)).

## Water supply

The supply of cooling water for all external grinding operations should be ample in volume but without force. A minimum flow of 25 litres per minute should be allowed for a 370 mm wheel and more in proportion for larger wheels. Used alone, water will rust the machine and work, but this may be prevented by adding $\frac{1}{4}$ kilogramme of common soda to each 5 litres of water. The white residue is more easily removed and the corrosive action of the soda minimised by adding, as well as the soda, about $\frac{1}{4}$ litre of machine oil to each 5 litres. Water, with a small proportion of soluble cutting oil (1 to 50), also makes a suitable cooling agent and is free from rusting and corrosive effects.

## The cylindrical grinding machine

This machine is made in two varieties—the plain and the universal types. The fundamental design is the same in each case, but whilst the plain machine is generally used for work between centres, the universal machine has more movements and is supplied with a greater range of accessories for increasing the range of work for which it is capable. In addition, it may be adapted for internal grinding. A diagram of a universal machine is shown at Fig. 250. The plain machine, although lacking the above refinements, is generally of more robust construction, and although more restricted in its range is capable of taking heavier cuts without loss in quality due to vibration, etc.

### Constructional details

**The bed.** The upper surface of the bed has the general form (in plan) of a ⊥ . Along the base of this are the ways for carrying the table, mounted with work-head and tailstock, whilst the wheel-head on its slide is carried on the vertical limb of the ⊥. Usually it is arranged that the machine base follows the three-cornered shape with respect to the points upon which it

*Landis Lund Ltd*

Fig. 250 A universal grinding machine (capacity 350 mm × 820 mm)

rests. This is an important aspect because, as the reader knows, a 3-legged stool always rests on its three legs whatever the shape of the surface upon which it stands. If four points of support are used there is always the possibility of one being unsupported, and the distortion that this might cause in the machine frame would be fatal to the accuracy of a precision-grinding machine. To avoid any possibility of such distortion grinding machines are often set down and levelled up, but not bolted.

**Wheel-head and slide** (Fig. 251). On the universal machine this is mounted on an intermediate turntable permitting it to be swivelled and fed at any angle. On the plain machine the head is fixed and moves perpendicular to the table. The head may be moved by a wheel placed in the front of the machine or, as in the type shown, by a wheel mounted on it. The movement ratio, wheel to head, is large, so that feed increments less than $\frac{1}{100}$ mm may be obtained and to eliminate the effects of backlash in the mechanism the movement is pre-loaded, either by a large weight on a chain or by other suitable means. By this means contact is always maintained on one side of the operating mechanism, and whichever way the operating wheel is turned the slide responds immediately without any lost motion.

**The spindle.** The wheel spindle is the most important unit in the construction and upon its efficiency depends the success of the machine to produce a ground surface of the requisite quality. Unless the spindle and its bearings are of the best design and workmanship the machine will never maintain a consistent output of precision ground work. The grinding head and spindle of a high-class universal grinding machine represents an exceptionally fine piece of machine tool construction. The grinding wheel must be held and rotated on a true axis, without any trace of vibration; otherwise it cannot produce work truly smooth and cylindrical. Until the development of modern bearings for this class of work, where the adjustment of the running fit is constant and controlled in some way, it was necessary to base the running fit clearance on the temperature conditions prevailing at the 'operating temperature' (about 60 °C) and, as far as possible, maintain the

*Landis Lund Ltd*

Fig. 251 Wheel-head of universal machine [note internal grinding spindle (above grinding wheel) and motor swung out of position]

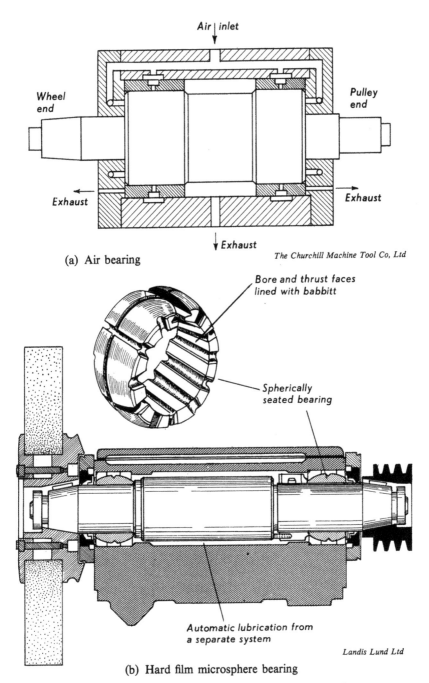

Air inlet

Wheel
end

Pulley
end

Exhaust

Exhaust

Exhaust

(a) Air bearing

*The Churchill Machine Tool Co, Ltd*

Bore and thrust faces
lined with babbitt

Spherically
seated bearing

Automatic lubrication from
a separate system

*Landis Lund Ltd*

(b) Hard film microsphere bearing

Fig. 252 Spindle bearing arrangements for grinding machines

spindle at that temperature by running it continuously. On the Churchill machines this problem is overcome by means of an air bearing, in which the journals and the thrust faces of the spindle float on a film of air maintained at a pressure of about 4·1 bar. By this means friction between the spindle and its housing is eliminated, no heat is generated and thermal effects can be disregarded. The system, of course, necessitates a supply of compressed air, with the ancillary regulators and filters, a diagram of this bearing is shown at Fig. 252(a).

The bearings on the Landis machine shown at Fig. 250 are shown at Fig. 252(b) and are known as hard film microsphere bearings. Here a large diameter spindle is supported in spherically seated bearings of special design and the interior of the housing is subjected to an automatic lubrication system with the oil at a pre-determined pressure. The makers of both these units claim that with normal care and maintenance the spindle and bearings will outlast the life of the machine to which they are fitted.

The belt drive to a grinding spindle has some influence on the results obtained and the drive should be soft and pliable so as to transmit the power evenly. This condition is satisfied by the use of endless rubber bonded vee belts.

**The table.** The table is carried on the ways at the top of the bed front and carries the work head and tailstock. Automatic lengthwise travel is provided for, the length of stroke and reversal being looked after by adjustable trips which operate the reversing lever. Most tables are now operated hydraulically (see later). Depending on the design of the machine, the table may be flat or triangular in section, the latter shape often being found on plain machines because, for this type of machine the extra rigidity is more important than the greater versatility of the flat table. The table is separate from the slide which carries it, being located and held by a pivot at the centre, clamping nuts at the ends and at one end, an adjusting screw for rotating it on the centre pivot. By this means the axis of the table (and centres) may be set at an angle for taper work and corrected for parallelism. A scale of degrees, taper per unit length, etc., is generally incorporated at the end where adjustment is made. The central pivot on which the upper table is rotated can be seen on Fig. 251 and the end of the table with the scale for setting is shown at Fig. 253, where the elements are designated.

**Work head.** The construction of the work head of a universal machine will be gathered from Fig. 260 which shows the head adapted for working with a live spindle. The internal construction allows for either the spindle

A. A. Jones & Shipman Co, Ltd

Fig. 253 Separate pivoting table-top of grinding machine
(see text)

to be driven in the same manner as a lathe with a chuck, or other work-holding attachment, fitted to its nose, or, when used with a centre, for the spindle to be locked and a front driving plate attached, which is driven independently and which carries a pin for driving the dog attached to the work between the centres (see also Fig. 247). For grinding sharp tapers on work held in the chuck the head may be swivelled to any angle, as shown by the circular base and scale on the diagram. The tailstock which supports the outer end of the work is similar to a lathe tailstock except that the barrel, instead of being screw-operated, is actuated by a spring. This allows it to accommodate for changes in work length caused by heat expansion, which would otherwise result in the work bowing in the centre and digging against the wheel.

**Work steadies**

The use of steadies is more important in grinding than in turning, owing to the greater accuracy and different cutting conditions involved. In addition to their usual function of supporting thin and slender work, steadies play an important part in suppressing vibration and so assisting the quality of finish and the wear of the wheel. When a shaft which is long in comparison with its diameter is being ground it is liable to be continually changing its axis as it revolves, and this tendency is aggravated if the heat caused by grinding is not dissipated as rapidly as it is generated. The steadies, by exerting some tension, prevent this and assist the work to revolve on a constant axis. Grinding steadies consist of two shoes which should bear on the work approximately in the relative positions shown by Fig. 254(a). The bottom support must not allow the work to fall towards the grinding wheel, but must tend to force it against the other shoe which should be a little above the centre. The net result of the action is that the shoes effect a wedging or braking action on the work, giving it steadiness whilst being ground. In order to compensate for the reduction of work diameter as it is ground, the back shoe is sometimes spring-fed. When applying the steadies it is not necessary to grind the point of application beforehand, but the steady should be applied immediately to the rough bar which, if out of round, will be brought to true roundness by the combined action of the steady and the wheel as grinding proceeds.

A diagram of a steady is shown at Fig. 254(b). The shoes which bear on the work are of wood, the lower one being adjusted by screw A and the upper one by C. In adjusting the steady, tension is first put on the work by A which is then locked. The block holder B is then fed forward by C to steady at the back of the work and then locked up by a screw behind (not shown). As reduction in diameter and block wear occurs on the upper block it is compensated by adjusting screw C forward. The unshaded portion of B represents a slot through which A passes.

The number and spacing of steadies for any particular ratio of length-diameter cannot be specified by rule as circumstances are so variable, but there should never be any hesitation to use steadies even although the work might appear large and rigid enough to be ground unsupported. For supporting the outer end of long work held at one end in the chuck, a three-point steady similar to a lathe steady may be used.

**Vibration and chatter**

These may be one of the most troublesome and elusive difficulties encountered on a grinding machine. Extreme cases of chatter, when the ridges on the work may be felt, are not likely to occur, or if they do are easily traced. It is when an extra fine finish is desired and chatter marks appear as delicate lines of light and shade, sometimes closely spaced and often travelling along the work in the shape of a helix, that the remedy is often difficult.

The two chief causes of this fault are connected with the wheel and the work. We have already discussed wheel balance and truing and if these, together with the belt and spindle condition are in order, attention must be given to the work. Owing sometimes to the regular spacing of the

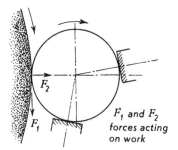

$F_1$ and $F_2$
forces acting
on work

(a) Disposition of steady shoes

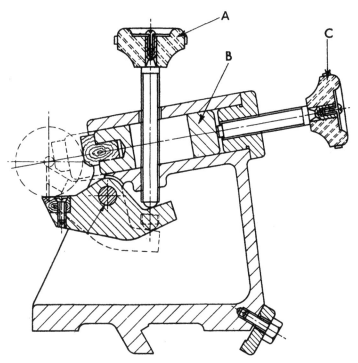

The Churchill Machine Tool Co, Ltd

(b) Grinding machine steady

Fig. 254 Steadying in grinding

marks they are often attributed to the presence of toothed gearing some-where in the machine, but it has been shown that they are just as likely to appear on work ground in a machine that has no gears at all in its con-struction. However heavy and rigid a piece of work may be, it will possess a certain natural frequency of oscillation governed by its weight and shape, and such work may be caused to vibrate by resonance from a movement of similar frequency occurring within the machine itself. The phenomenon is similar to that which will cause a piano string to vibrate when the note given by the string is whistled into the instrument. By using steadies, and experimenting with their positions along the work, chatter caused by the shape or dimensions of the work may be minimised or eliminated. Many cases of chatter and inaccuracy may be traced to the support of the job. A loose work-head spindle, when this is used as a 'live' spindle, may cause it or the driving dog when grinding on centres may give rise to vibrations. Some of the most obscure effects when grinding on centres may be traced to the centre holes, and it may be taken that the surfaces being ground can-not be more accurate than those which are providing the support. Centre holes should be round, clean and free from hardening scale and for ordinary work should be cleaned out with emery cloth. When extra accuracy is re-quired they should be lapped out by rotating a 60° conical oilstone in the spindle of a lathe, supporting the work with one centre hole against the tailstock centre and feeding the centre hole to be lapped against the oilstone.

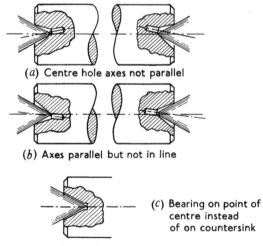

(a) Centre hole axes not parallel

(b) Axes parallel but not in line

(c) Bearing on point of centre instead of on countersink

Fig. 255 Faults in centre holes

Additional faults which may exist in connection with the centre holes are shown at Fig. 255. At (a) and (b) the axes of the holes are not in alignment, whilst at (c) the centre is bearing on its point instead of supporting on the countersink. All these faults may cause chatter, as well as inaccuracy in the finished surface.

**Operating the machine**

When the centre holes have been cleaned out and the driving dog fitted the tailstock should be adjusted to such a position that there is a reasonable amount of tension on the spring with the work in position. The original turning should be checked for grinding allowance and parallelism for guidance during the preliminary passages of the wheel. With the work in position the wheel may be fed to within 2 mm to 3 mm of its surface and a good stream of cooling water turned on. The work is now fed from end to end with the left hand, whilst the wheel is slowly fed in with the right, and ultimately the appearance of sparks will indicate that the wheel is cutting. Without adding any more cut the work should be fed across the wheel for its whole length, observation being kept as to whether the cut is running off or getting deeper. If the latter takes place it may be necessary to take off some of the cut and adjust the table, but if the original turning was reasonably accurate and the grinding does not seem to deviate much from it, a little more cut may be put on until the diameter has cleaned up sufficiently for a micrometer to be tried on it.

A check must now be made for parallelism and if this is at fault the table adjusted a small amount in the proper direction. (Wind the wheel away first if the adjustment will move the work nearer the wheel.) After adjustment another small cut should be taken along the bar, noting whether extra stock is being removed from the larger end, and when the sparks indicate that the wheel pressure is even over the whole length a further trial for size and parallelism must be made. This should be repeated until the work is parallel, and then it may be ground down to size with the automatic feeds, with the last hundredth or so of cut being hand-fed. If there are a number of similar pieces the work will be fairly rapid after the first one, as the table will be set for parallelism and the trip for the ratchet operating the in-feed may be set to leave off feeding when the work is almost on size, the last fraction being removed by hand-feed. If the amount left on for grinding is too small to risk getting undersize during the preliminary setting of the table, a piece of scrap material the same length should be put in and ground up for table setting.

### Grinding to a shoulder

When work has to be ground up to a shoulder it should be undercut so that the wheel may grind the diameter without touching the shoulder (see Fig. 99(a)). It cannot be expected that the table reversing trips will stop the table to the same exact line each time, so that they should be set to reverse the table when it is 2 mm to 3 mm from the shoulder. The small amount of metal (if any) left after the remainder of the diameter is finished may be ground off afterwards by setting the wheel close to the shoulder and feeding it in. When a number of similar jobs have to be ground up to a shoulder, although a measurement check may prove the turned lengths to be uniform, the reversal of the grinding wheel takes place at a constant distance from the headstock *centre*. If the centre hole sizes vary, therefore, the setting for one may not be suitable for the next, and care should be taken that the wheel is not allowed to crash into the shoulder. For special face grinding work where it is necessary to obtain accurate longitudinal positions of the table, and to repeat the same position, dial indicators and stops may be mounted. This feature is shown at Fig. 256. By setting one indicator against its stop and then inserting a length bar or stick micrometer and re-setting, an accurate movement of the table may be made.

*A. A. Jones & Shipman, Ltd*
Fig. 256 Longitudinal position indicators for table

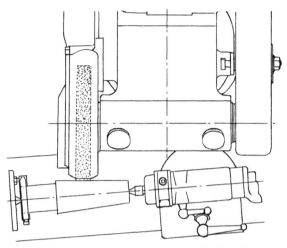

*The Churchill Machine Tool Co, Ltd*

Fig. 257 Grinding taper on universal machine

## Taper grinding

Slow tapers are ground by swivelling the machine table to the required angle (Fig. 257), whilst tapers having an angle beyond this capacity may be dealt with on the universal machine by swinging and feeding the wheel-

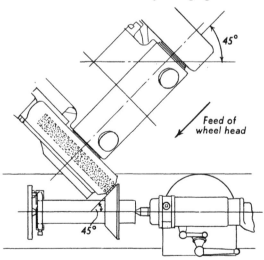

*The Churchill Machine Tool Co, Ltd*

Fig. 258 Grinding angular face on universal machine

head slide at the proper angle (Fig. 258). When the wheel-head is twisted the wheel itself should be adjusted and trimmed so that its face is parallel with the direction in which it moves. Sharp tapers, such as the 60° points of centres, may be ground on work held in the work-head chuck or spindle by swivelling the work-head to the required angle, and grinding with the face of the wheel in its normal position.

**Facing.** As we have already discussed, work on centres may be faced on its ends and shoulders by using the side of the wheel. Working this way, the shape of the face is governed by the relative wheel-work alignment. If the axes of the wheel and work are parallel the face will be flat, but if there is any deviation the face will be humped or hollow, depending on the direction in which the axes are out of parallel. A face should either be flat or slightly hollow. It should never be humped unless special instructions have been given to make it so. On the universal machine, facing may also be carried out on chuck or faceplate work by using the internal grinding head upon which is mounted a flaring cup wheel (Fig. 264(b)). Another facing method

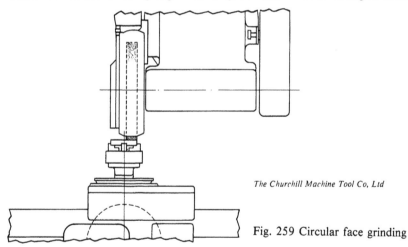

*The Churchill Machine Tool Co, Ltd*

Fig. 259 Circular face grinding

is shown at Fig. 259 where the work-head is swivelled through 90° and the grinding performed by the face of the wheel. In each case a test for flatness should be made before the finishing cut is taken over. This can be done with a straight-edge or with the wheel itself by feeding it beyond the centre. If the facing is flat the amount of sparking should indicate the same wheel pressure when it is fed on to the face beyond the centre from the side on which the face was ground. The reader may think out for himself the condition of the face when the wheel clears or alternatively digs deeper into the other side.

# 10  Grinding (*continued*)

## Internal grinding

The universal machine may be adapted for internal grinding by bringing a special spindle into position. On the Landis machine shown at Fig. 250 the internal spindle is mounted at the side of the wheel-head and brought into position by swinging it down, the working arrangement being shown

*Landis Lund, Ltd*

Fig. 260 Arrangement of universal machine for internal grinding

at Fig. 260. On other well known makes of machine this spindle is placed at the rear of the wheel-head and brought into position by swivelling the head through 180°. It is the general practice, however, to concentrate quantities of internal grinding on an internal grinding machine, an

*Rockwell Machine Tool Co, Ltd*

Fig. 261 Internal grinding machine

example of which is shown at Fig. 261. The machine incorporates a table with work-head which carries a chuck or other suitable means for holding the work. At the other end of the machine is the wheel spindle, mounted, with its driving arrangements, on a cross-slide. The machine shown is of a semi-automatic type having a capacity for grinding up to 40 mm bore and 70 mm length. The movements may be manually operated, but when the machine is set for the semi-automatic cycle, after a bush has been loaded, grinding of the bore takes place automatically until the limit of the roughing cut has been reached. The wheel then disengages from its work and must be trued for the finishing cut. The automatic cycle is then renewed and the wheel again disengages when the finished size is reached. A truing diamond arrangement is built into the machine and is carried in the inclined holder seen in front of the guarded chuck. The vertical unit to the left of the wheel-head is a facing attachment which may be swung into position for facing work with a cup wheel (see Fig. 264(b)). Unfortunately in the illustration the actual internal grinding wheel is hidden by a safety cover provided to swing in front of the wheel. This is necessary because in internal grinding there is the hazard, when trying the hole with a plug gauge, of the gauge suddenly pulling out, resulting in the operator's hand hitting the wheel unless this is guarded.

**The spindle.** The operating conditions imposed by this class of work, involving a small wheel and its consequent high rate of rev/min, together with the fact that part of the spindle must enter the hole being ground, renders the use of a special spindle necessary. Internal grinding spindles are of two main types: (1) the tube type and (2) the extension type. The tube type of spindle, which is the older in design, has the sleeve carried right up to the wheel, the end near the wheel carrying a bearing (Fig. 262(a)).

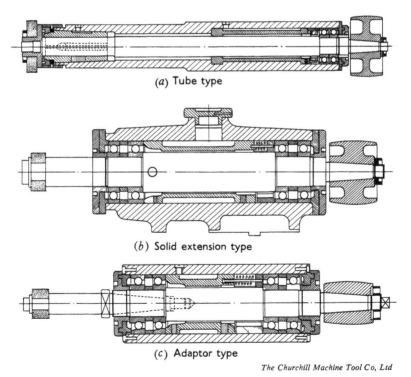

(*a*) Tube type

(*b*) Solid extension type

(*c*) Adaptor type

*The Churchill Machine Tool Co, Ltd*

Fig. 262 Internal grinding spindles

This has advantages in rigidity and freedom from deflection, but places a limit on the smallest hole that may be ground. For this reason the application of the tube type of spindle is mostly on large and long holes. On the solid extension spindle the wheel is carried at the end of an extension of the spindle, the wheel and extension entering the hole being ground. This type is also made with the extension in the form of an adaptor with a taper shank, which allows an adaptor to be used proportionate in size to the hole being ground. Solid extension and adaptor type spindles are shown at Fig. 262(b) and (c), where it will be seen that the adaptor fits with a taper shank and screwed end.

**Speeds and driving.** Although the peripheral speeds used for internal grinding are less than for other classes, the small wheels employed necessitate high revolutions and special arrangements for obtaining the necessary condition. For example, a 25 mm wheel to operate at 750 metres per minute surface speed would require to rotate at $\dfrac{1000 \times 750}{\pi \times 25} = 9550$ rev/min. The high speed and the shortness of the belt necessitates that it should be made endless, and of a composition able to work under the conditions imposed.

**Other machine details.** The work-head of the internal machine is on the same lines as that for the universal except that it is permanently employed with a live spindle. The table generally has provision for swivelling its upper portion about the centre as on the external machine. On the larger machines automatic feed is provided, but on small and medium types where the length of bush to be ground is not likely to be great, the travel of the table is hand-operated. The cross (wheel-head) slide of the internal machine is not generally provided with such a fine and delicate feed as the corresponding unit of the external machine because, on internal work, using spindles which are often far from robust, spring takes place which does not allow a cut to work itself out. The operation, therefore, on such work is completed under the guidance of the operator's experience, aided by the intelligent use of the plug gauge rather than by relying on the application of so many known increments of wheel in-feed to a final stopping point. Production type internal machines are made to operate on an automatic cycle in which gauging, wheel truing, etc., take place unaided by the operator, and once set and loaded the machine grinds a bore to size without any attention (see description of Fig. 261).

## Internal wheels

The wheel diameter for internal work is governed by the size of hole to be ground, and as a rough guide the following diameters should be used:

| Hole diameter (mm) . . . . | 150 | 100 | 50 | 25 | 13 |
|---|---|---|---|---|---|
| Wheel diameter (mm) . . . | 100 | 75 | 38–44 | 19–22 | 9–11 |

The larger wheels will be new wheels made to the correct size and about 19 mm to 20 mm wide, but the small ones may be made from broken pieces of other wheels, drilled for the hole, roughly shaped and then trimmed circular on the spindle.

According to the rules governing wheel selection the large arc of contact occurring with internal work calls for a soft-grade wheel, and whilst this condition may be satisfied when working with the larger wheels, small ones often lack the necessary mechanical strength to withstand breaking and must be used in a harder bond than is really suited to the work. For this reason there is often a tendency towards glazing which necessitates frequent dressing of the wheel.

## Work setting

Work speeds for internal grinding may be about 50% higher than for external work and as a basis for setting a speed of about 30 metres per minute may be used. When setting up the work, surfaces which may have subsequently to be ground must be set true or, if grinding the hole is the only finishing operation, the work must be set with other important surfaces true. For example, a shouldered bush to be ground later on the shank and shoulder would be set with these surfaces true. This may be done with a clock indicator, but after a few weeks of practice results just as effective may be obtained much quicker with the help of a piece of chalk held in the hand. If the grinding of the hole is the only operation, the work will have to be set up according to circumstances. For example, if setting up a gear for grinding out the bore, it would be necessary to ensure that the sides of the gear and the teeth were set true. This could be effected by strapping the gear to a true faceplate and resting four or five similar and suitably sized test plugs in the tooth spaces at equidistant points round the circumference. By passing a wire round the gear the plugs could be held in their positions and then they could be set true with a dial indicator.

## Machine operation

When grinding large bores with a reasonably large wheel and robust spindle the operation is not affected unduly by deflection, and if the machine is correctly set parallelism is not difficult to obtain. With smaller and less rigid spindles, however, the wheel tends to be pushed away when cutting on its full width and to cut more at the ends where a portion of its length leaves the hole (about $\frac{1}{3}$ length as for external grinding), resulting in a tendency for holes to be ground bell-mouthed. Since most of the smaller machines are hand-operated the grinder, as he gains experience and sympathy with the process, instinctively allows the wheel to dwell longer at the region of the hole he knows will require additional grinding to counteract the tendency for the ends to be large, and so obtains a parallel hole. In this he is helped by the feel of the plug gauge during the final stages of sizing and when doing this operation readers are warned, when the gauge enters a short distance in the hole of a bush which is warm from grinding, not to allow it to stay too long or it will be necessary to drive it out and spoil the setting of the bush. The machine shown at Fig. 261 has a safety guard over the wheel, to protect the hand of the operator when a tight plug gauge might suddenly respond to being pulled.

The rate of longitudinal table travel for internal grinding may be fairly high, as this helps to spread the heating effect since the operation is generally performed without a liquid coolant. If the travel per work revolution is equal to or greater than the wheel width (and it might be greater in hand operation where accurate judgement is impossible) it does not matter, as when a cut has been applied many passes of the wheel will probably be necessary before the spring has worked itself out and allowed the spindle to resume its normal position.

## Operation sequence

For bushes and round work generally, internal grinding nearly always takes place before any external finishing, which is done with the work on a mandril. The grinding of faces will depend for method on the proportions of the surface and the facilities available for doing the job. The following examples will serve to illustrate how the grinding would be arranged on the components in question:

*Example* 1. *To grind the bush shown at Fig. 263(a)*

1. Hold in 4-jaw chuck. Set up true on shank. Internal grind the bore (b).

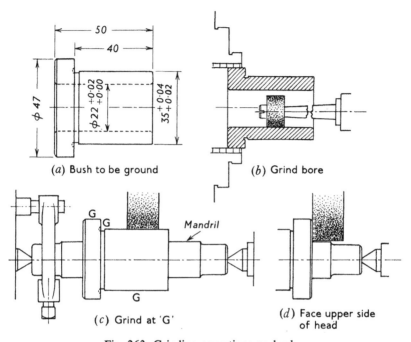

(a) Bush to be ground

(b) Grind bore

(c) Grind at 'G'

(d) Face upper side of head

Fig. 263. Grinding operations on bush

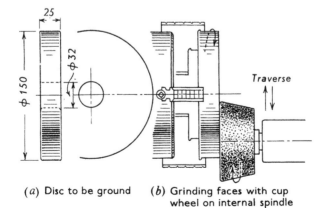

(a) Disc to be ground

(b) Grinding faces with cup wheel on internal spindle

Fig. 264

2. Press on mandril and set between centres of cylindrical grinding machine.

Grind shank and top diameter of head with face of wheel and underside of head with side of wheel (c).

3. Reverse mandril in machine. Grind upper surface of head (d).

*Example* 2. *To grind the component shown at Fig. 264(a)*

1. Grind the sides flat and parallel. This can be done either by (*a*) the universal machine by the method shown at Fig. 259, using a jaw chuck or rotating magnetic chuck, or (*b*) with the universal or internal machine with a cup wheel as shown at Fig. 264(b).

2. Set up on the faceplate of an internal grinder. First clamp with a long bolt through the hollow spindle and set the top diameter true, then hold with three finger clamps, remove the centre clamp and grind out the hole.

3. Press on to a mandril and grind the top diameter.

## Surface grinding

Surface grinding has made rapid advances not only in its scope as a finishing operation, but also as a machining process for finishing surfaces from the rough, thus eliminating planing, shaping or milling. One of the important applications to which it has been put is that of finishing the slideways of machine-tool beds and so obviating the necessity of hand scraping. As a process for finishing from the rough it has been used widely on the joint faces of castings of the type used for engine cylinder blocks, cylinder heads, etc.

Flat surfaces may be ground either by using the periphery of a disc wheel or by grinding with the end of a cup (cylinder) wheel. These two fundamental methods may be further subdivided according to the method adopted on different types of machines for feeding the work relative to the wheel.

### 1. Disc wheel

(*a*) *Horizontal spindle—reciprocating work* (Fig. 265(a)). In principle this is similar to shaping and planing, and is used mainly for light precision work in connection with tool and gauge making.

(*b*) *Horizontal spindle—rotating work* (Fig. 265(b)). The basis of this method is the same as that for facing on the lathe. It is used for facing

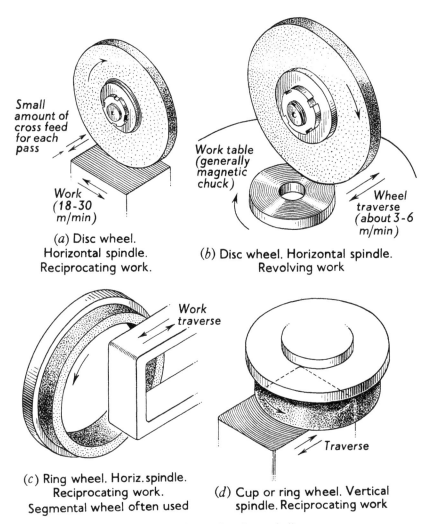

Small amount of cross feed for each pass

Work (18-30 m/min)

(a) Disc wheel.
Horizontal spindle.
Reciprocating work.

Work table (generally magnetic chuck)

Wheel traverse (about 3-6 m/min)

(b) Disc wheel. Horizontal spindle.
Revolving work

Work traverse

(c) Ring wheel. Horiz. spindle.
Reciprocating work.
Segmental wheel often used

Traverse

(d) Cup or ring wheel. Vertical
spindle. Reciprocating work

Fig. 265 Methods of surface grinding

circular work and narrow rings (e.g. piston rings), which are more easily and safely held on a rotating magnetic chuck. The finish given is circular, concentric with the work, and for some classes of work this appearance, although not essential, is desirable. (Readers may have noticed the effect of different appearances of finish. For example, a circular face finished with crosswise markings, however accurate, never has the appearance of accuracy such as would be given if the finish marks were circular with the work.)

## 2. Cup (ring) wheel

(a) *Horizontal spindle—reciprocating work* (Fig. 265(c)). The cup wheel produces a flat surface on the same principle as the face mill, and this method of grinding is similar to milling by the method shown at Fig. 145(b).

(b) *Vertical spindle—reciprocating work* (Fig. 265(d)). This method is similar to that of a face mill operating on a vertical milling machine. In design, the grinding machines using this method are of two main types: (i) open at the front, similar to the knee type vertical milling machine, and (ii) similar to the planing machine with the wheel spindle carried on a cross structure.

(c) *Vertical spindle—rotating work table.* The diagram for this is similar to Fig. 265 (d), except that the work passes under the wheel with a circular instead of a straight-line motion.

Of the five arrangements, 1(a) and 2(b) are by far the most common, being applied to grinding of all types of plane work. The general distinction between the two is that 1(a), whilst being a slower method of metal removal, is more versatile than 2(b), and permits faces to be ground up to shoulders, vertical faces with the side of the wheel, angular faces, and so on. For this reason it is used for tools, gauges and general toolroom work where a wide range is desired. Method 2(b) is a fast and accurate method of grinding plain flat surfaces, and this constitutes its chief, if not its only application.

Next in popularity comes method 1(b), and machines employing this principle find a useful place in many general grinding shops. It is used for grinding the faces of disc and thin round work such as circular saws, piston rings, etc. Machines employing methods 2(a) and 2(c) have been developed mostly on specialised lines for production work, particularly for grinding the surfaces of castings and forgings direct from the rough.

## Wheels

Closer attention is necessary in the selection of the grit and grade of wheels for surface grinding than in other classes of grinding, as unsuitable wheels are liable very quickly to cause local heating, blueing, surface cracks and other defects. This is because there is a considerably greater area of contact between the wheel and the work than for cylindrical grinding, and in the case of cup wheels the area of contact is so spread out that efficient cooling with water is difficult. The general rule for surface grinding, therefore, and particularly when using cup wheels is to use wheels of a softer grade and coarser grit than for cylindrical work.

When disc and flanged cup wheels are used they are mounted between flanges in the same way as on any other machine, but the hollow ring type of wheels has to be cemented to the adaptor which fits on the end of the spindle. This cementing may be carried out with Portland cement, fusible cement or molten sulphur. Fusible cement or sulphur possesses the advantage of quick setting, as the time required for Portland cement to harden is about 48 hours. When mounting a new wheel, the adaptor should be cleaned of the former wheel and cement, and the new wheel placed centrally in it. The securing compound, in a thick liquid state, is then poured in and allowed to set. If the wheel is provided with wire reinforcing bands these should be towards the projecting portion of the wheel and away from the adaptor. The mixture of Portland cement suitable is equal parts of cement and sand. When a fusible compound is used on no account must the steel adaptor be heated as the wheel may be fractured when it cools and contracts.

The larger cup wheels are made up of segments which are held in a special chuck (Fig. 234). These have two principal advantages over a continuous ring wheel: (1) if a segment is broken it is cheaper and easier to replace than a complete wheel, (2) the spaces between the segments help to keep the wheel face clear and promote freer cutting conditions than is the case with the continuous cutting face of the ring wheel. The ring wheel, however, is more suitable for thin, frail work and where finish or accuracy are important.

**Wheel truing**

The method to be used for keeping the cutting face of the wheel in condition will depend on the class of work upon which it is operating. For important work and when grinding is being done by the periphery of a disc wheel a diamond should be used. For surface grinders these may be set in a special holder which may be held in the vise or on the magnetic chuck (Fig. 266). When a cup wheel is engaged on comparatively heavy roughing cuts, it is more important that its cutting face should be open and rough than that it should be perfectly true. It may be kept in this condition with a dresser, and in such cases it would be exposing a diamond to unnecessary wear to use it. The cup wheel tends in general to keep its face true, but periodic dressing or

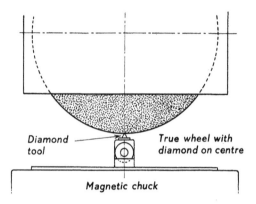

Diamond tool

True wheel with diamond on centre

Magnetic chuck

Fig. 266 Wheel truing in surface grinding machine

truing will be necessary to remove loaded metal or glazed abrasive grains.

## Work holding

The most usual methods of holding work on the surface grinder are as follows: (1) By clamping direct to the table, to an angle plate or some such fundamental set-up; (2) Holding in a vise; (3) Holding in some form of special fixture; (4) Magnetic chuck.

The first three methods are used in cases where the grinding to be done is not of the form in which a face has to be ground parallel with another face, both being external and of reasonable dimensions. For ferrous work on which this last condition holds good the magnetic chuck is employed, and as a large proportion of the output of surface grinders is of this type the magnetic chuck is a common auxiliary to the machine. Some machines have such a chuck incorporated in the design as the permanent table, and if a vise or fixture has to be used it must be rested on and held to this chuck by the magnetic force.

## The vise

The general design of vises used on surface grinders is on the same lines as for milling machines, except that they are generally rather smaller and not so commonly equipped with a swivelling base. Rigidity and freedom from vibration are important so that grinding vises are as low and squat as possible. For angular work the universal vise shown at Fig. 57(c) is useful.

## The magnetic chuck

Magnetic chucks are made in circular and rectangular form in sizes ranging from approximately 150 millimetres to 2 metres diameter for the circular, and 150 millimetres to 2 metres long in the rectangular shapes. The smaller sizes of such chucks may be obtained either with permanent magnets or as the electromagnetic type.

*Electromagnetic chucks.* These use the principle of electromagnetism by which a bar of iron, when enclosed in a coil of wire, is magnetised when an electric current is passed through the coil, its end having N and S polarity according to the direction of current flow in the coil. In its construction the body or base of the chuck is incorporated with pole pieces of special high permeability magnet steel. These pole pieces are surrounded with coils of insulated copper wire for carrying the electric current, the direction of winding being arranged so that the upper ends of the magnets have alternate N and S polarity. The top plate or pole face of the chuck is made up of alternate steel, and non-magnetic segments, the object being to break up the magnetic continuity so that the magnetic lines of force, instead of finding a path through the plate, are compelled to emerge from its upper surface and pass through the object requiring to be clamped. A diagram of a chuck, showing the direction of the magnetic flux, is shown at Fig. 267.

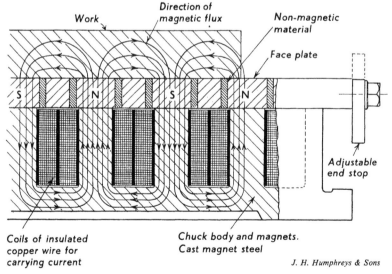

Fig. 267 Section of electromagnetic chuck

The current to these chucks must be controlled by a special switch, and one of these is shown at Fig. 268. The chuck may be switched on by moving the handle to either side, and when switching off, if it is reversed to the opposite side first, the reversal of current effecting a more complete demagnetisation of the poles than merely disconnecting the current. If it is necessary to hold a component on the chuck with packing between the work and chuck face

*J. H. Humphreys & Sons*

Fig. 268 Magnetic chuck (tilting pattern) with operating switch

a packing of solid, continuous steel is useless, as it absorbs the whole of the magnetic flux leaving none emerging for the purpose of holding the upper piece. By using a packing made up of laminations of magnetic and non-magnetic material (e.g. steel and brass) some of the magnetic flux emerges and is available for holding work either to the upper surface or sides of the block (Fig. 269). Electromagnetic chucks must be supplied from a direct current supply and are usually made for voltages of 100–110 and 200–250.

*The permanent magnet chuck.* The general shape, construction and appearance of this chuck is similar to the electric type, the principal difference being that the magnets, instead of deriving their magnetism from an electric current, are permanently magnetised. For this purpose they are made of a special alloy steel having the property of taking on a high stage of magnetisation and retaining it over a long period. In the 'Eclipse' chuck the magnets are assembled in a silicon iron grid to form a magnetic 'pack'. This grid forms the power unit and the chuck is made up by enclosing the grid in a non-magnetic metal casing which is assembled between base and top plates, both of dead mild steel, the top plate having inserts of Armco iron separated magnetically from the steel plate by a layer of white metal.

THE MAGNETIC CHUCK 359

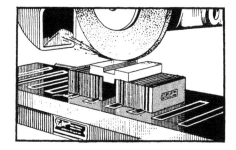

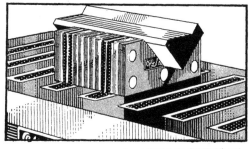

*James Neill & Co (Sheffield) Ltd*

Fig. 269 Holding irregular work via laminated packings

The grid has a small longitudinal movement controlled by an eccentric shaft operating through a link. The above details may be followed from Fig. 270 which shows the dis-assembled units of the chuck.

An interesting feature of the chuck is the method of switching it on and off, and this is achieved by a small movement of the grid. When this is in the 'on' position the solid members of the grid line up with the pole inserts of the top plate (Fig. 270(b)). In this position the magnetic flux follows the path shown dotted, i.e. S into outer pole (top plate), N into baseplate, through the grid into the inner pole (insert), the circuit being completed by passing through the work. In the 'off' position (Fig. 270(a)) the magnetic pack has moved so that the grid and pole inserts are out of line. The path of the magnetic flux now is still through the baseplate and grid, but the circuit is closed through the top plate and inserts instead of through the workpiece.

A very useful chuck in the 'Eclipse' range is the 'Minor', a small portable article with working surfaces $125 \times 62$ mm, having a mass of about 4 kg. Both upper and lower faces may be magnetised, and the switch shown on the end is arranged so that the top, or bottom, or both, may be energised.

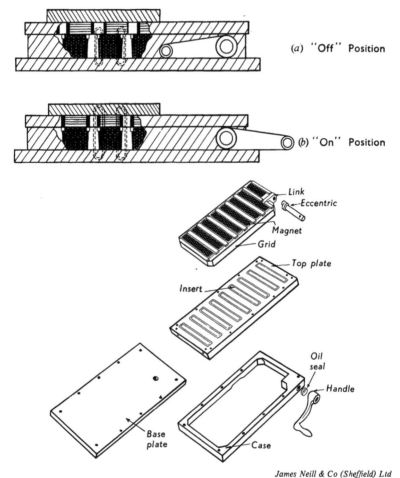

(a) "Off" Position

(b) "On" Position

Link
Eccentric
Magnet
Grid
Top plate
Insert
Oil seal
Handle
Base plate
Case

James Neill & Co (Sheffield) Ltd

Fig. 270 Details of 'Eclipse' permanent magnet chuck

(*Above:* Assembled chuck in 'off' and 'on' positions.)

This chuck is useful not only for grinding operations but also for marking out, filing, supporting a dial indicator and the hundred and one other cases where thin fragile articles must be held (Fig. 271).

In comparing the relative merits of the two types the permanent magnet chuck is independent of a d.c. electric supply, and of the special generation or rectification apparatus necessary in a plant supplied only with a.c. Since the chuck is self-contained, without any leads to an electric supply, it is more portable and may be set up on any machine where it might be useful. In favour of the electric type there is practically no limit to the size or shape to which a chuck can be constructed, and for certain machines a circular chuck may form the permanent table. Provided the supply voltage is kept within 10% of its correct value the holding power of the

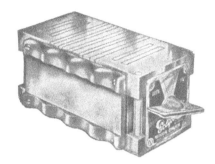

James Neill & Co (Sheffield) Ltd

Fig. 271 'Eclipse' minor chuck

chuck will not vary. The best permanent magnets tend to lose some of their magnetism, with time, particularly if subjected to shock or vibration, and the power of such a chuck may fall after some years of service necessitating re-magnetising of the magnets.

**Demagnetisation.** After a component has been on a magnetic chuck it is necessary to demagnetise it as the residual magnetism may be detrimental in service. Hardened steel tends to hold its magnetism more than soft steel and cast iron, but all metals should be treated. The operation is performed by subjecting the article to an alternating magnetic field which quickly destroys any residual magnetism, and a demagnetiser consists of a powerful electromagnet energised by an alternating current supply. On the *platen* type magnetised articles are placed on the platen and so become influenced by the field of the magnet located underneath. The *armature* type demagnetiser has two external laminated magnetic poles and articles may be treated by passing them through the space between. This type is useful for dealing with work which cannot conveniently be placed on a flat plate.

### Surface grinding machines

Probably the oldest established design of surface grinder is that which employs the edge of a disc wheel mounted above a reciprocating table. The design of machines of this type has been cleaned up considerably in recent years and a modern example of it is illustrated at Fig. 272. The machine shown has a table working surface measuring 457 mm × 152 mm and the

Fig. 272 Periphery wheel surface grinder

longitudinal and cross traverse of the table may be by hand, or by hydraulic operation. The hydraulic drive is achieved by a pump unit mounted on the floor inside the machine.

The spindle, running in anti-friction bearings, carries a grinding wheel 178 mm diameter and is driven by an endless, flat belt from a motor mounted inside the column. Using the maximum (178 mm) wheel, the greatest height that may be ground is 240 mm from the table surface.

With the development of alternative and more robust methods of grinding the applications of this type of machine are now confined almost exclusively to light precision work where accuracy, finish and adaptability are the chief requirements. It is essentially a 'sensitive' machine, and as such is suited to the finishing of important tools and gauges for which work its rather high, open front enables the operator to make settings and follow the progress of the grinding without any difficulty. When being used on a full range of toolroom work a good deal of grinding will be done by using the side of the wheel as well as its face. This enables a vertical face to be ground at the same setting as a horizontal surface and necessitates preparing the side of the wheel in a manner similar to that we have discussed for facing shoulders on the cylindrical machine.

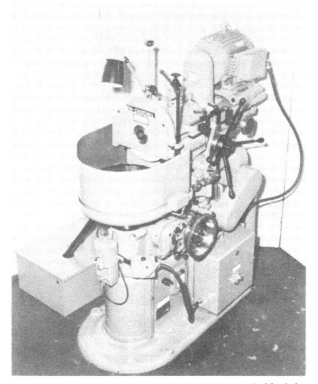

*The Churchill Machine Tool Co, Ltd*

Fig. 273 Periphery wheel rotating table surface grinder

An example of a machine employing the edge of a disc wheel mounted above a rotating work-table is shown at Fig. 273. On the machine illustrated the work-table is a magnetic chuck, and although the machine can be optionally supplied with a plain table, the nature of the work is generally such that a chuck is the more useful. The wheel spindle, driven by vee belts from the motor on top, is carried, with the motor, on a slide which moves at speeds of 1·17 or 1·7 m/min, and provides sufficient travel for the wheel to operate between the centre and the rim of the chuck. The machine shown has a 305 mm chuck so that the maximum stroke of the wheel slide would be approximately 155 mm. The table is provided with three speeds of 54, 95 and 150 rev/min from a variable speed motor and is elevated by the large hand-wheel, or automatically by ratchet, with a feed of 0·0025 mm per tooth of the ratchet wheel. For grinding parallel surfaces the wheel slide is set horizontal, but provision is made for tilting it 3° on either side of the horizontal plane to enable convex or concave surfaces to be ground. This feature is useful when components such as slitting saws are being ground on their side faces because the table may be set to grind a few thousandths of concavity, a necessary feature if the body of the saw is to clear the slit it is producing. The scope of this type of machine is not limited to circular components and it can be applied to the surface grinding of any shape that can be contained on its chuck. The machine provides a very good method of grinding the faces of blanks, piston rings, milling cutters etc., where circular, concentric finish marks are preferred.

The third type of machine we propose to illustrate is the most usual

*The Churchill Machine Tool Co, Ltd*

Fig. 274 Vertical spindle ring wheel surface grinder

variety of machine which employs a cup or ring wheel on a vertical spindle, located above a reciprocating table. On the machine shown at Fig. 274 the spindle is the armature shaft of the 30 kW (960 rev/min) driving motor mounted vertically, and at its lower end carries a flange upon which a ring wheel is mounted. The table traverse is hydraulically operated and permits a stepless range of speeds up to 30·5 m/min. On the particular machine illustrated the wheel is segmental, 457 mm diameter, and is large enough to cover the widest work accommodated, the working surface of the table being 420 mm wide. On some machines of this class a cross-feed is incorporated enabling a surface to be covered which is wider than the wheel. The cross-feed feature is often also useful in permitting a flat surface to be produced when the wheel axis is slightly tilted from the vertical. This is

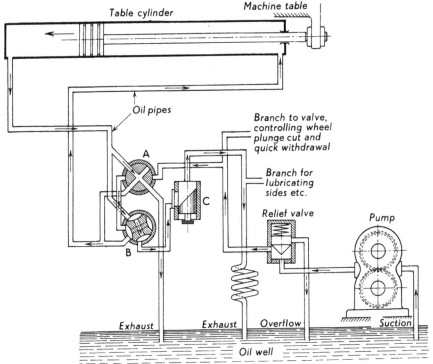

Fig. 275 Hydraulic circuit for operating grinding machine table

*A*, Stop and start valve (shown in stop position); *B*, Table reverse valve (shown for pressure to RH end of cylinder); *C*, Speed control valve (operates by varying the back pressure)

often done to reduce the area of wheel contact and minimise heating, but unless the wheel is traversed across the surface being ground a slightly concave effect will be produced. In Fig. 274 a magnetic chuck may be seen bolted to the machine table.

**Hydraulic driving**

All the grinding machines we have shown have their tables traversed by hydraulic means, and this method, developed about forty years ago, is now applied to the operation of most grinding machines where continuous operation of the table is necessary, and on some machines it is used for other functions as well. It is also used to drive the tables and other movements of machine tools other than grinders, although its application to the latter is most common. In principle it is very simple—in fact, its simplicity is one of the factors in its favour. Underneath the table, or at a suitable position relative to the slide to be operated, is a long cylinder fitted with a piston which is coupled to the table or slide to be moved. By arranging for oil under pressure from a pump to be fed alternatively to each end of the cylinder the piston and table are caused to reciprocate to and fro. The trips at the front of the table, through the reversing lever, operate valves which direct the flow of oil and another control is fitted which varies or restricts the amount of oil delivered and by this means enables the speed of the piston to be adjusted to any speed from zero to a maximum. A diagram showing the oil circuit for traversing the table of a grinding machine is shown at Fig. 275.

Before the application of this form of operation, table and slide movements were obtained in the usual way by gears, rack and pinion, leadscrews, etc. Some of the advantages of the oil method are:

(a) A stepless range of speeds is possible from zero to the maximum. With a mechanical method the number of speeds is limited to the number of gear changes available unless the drive is by variable speed motor.

(b) Much higher speeds are possible.

(c) Instant reversal without pause or shock at the ends of the stroke.

(d) Simpler control and fewer working parts to wear and develop slackness.

(e) The system has a greater element of safety since, if a jam or solid obstruction occurs, the oil will by-pass through the relief valve and no damage will be done. A lock on a mechanically operated slide generally results in severe damage to the mechanism.

(*f*) Smoother drive and hence less risk of gear vibrations affecting surface finish.

Grinding machines of the cup wheel type have already been illustrated, and other designs employing the same principle of operation (e.g. planer type, rotating table, etc.) are used for the straightforward production of flat surfaces, particularly on work where there are two surfaces opposite, enabling the support to be taken on one face whilst the opposite one is being ground. For such work the cup or ring wheel used in conjunction with the magnetic chuck forms an ideal combination, as for the rapid removal of material cup, ring or segmental wheels are well in advance of the method using the periphery of a disc wheel. This advantage is offset by the facility with which surfaces may be ground up to shoulders, etc., with the disc wheel, so that in a shop where a considerable amount of surface grinding has to be done there should be a machine of each type if the work is to be carried out with maximum efficiency. As we shall see later, the universal tool and cutter grinder, as well as its own function, lends itself to light surface grinding with the disc wheel and one of these, together with a machine after the style of Fig. 274, forms a good combination for surface grinding in a small shop. Naturally, where the volume of work is larger, more machines in greater variety may be installed to take advantage of the particular qualities of each machine.

**Surface grinding procedure**

In grinding, more than on any other operation, extreme care should always be taken not to distort the work in any way during clamping. The cutting forces are much less than those applied in milling or shaping so that moderate pressures are sufficient to hold the work. When using the magnetic chuck it should be remembered that the chuck may be relied upon to hold the work down, but additional stops may be necessary to prevent it from being pushed along the face of the chuck by the side thrust of the wheel. If the face to be held on the chuck is rough and irregular the opposite face should be smoothed up with light cuts and the work turned over before any attempt is made to remove much material.

When the work has been set up the wheel should be lowered as near to it as is possible by eye and then, if there is a considerable length to be ground, the work should be traversed to and fro over its whole length, and the wheel gradually brought down, attention being given for the first sign of sparks indicating that contact has taken place. The automatic feed may then be engaged and a few increments of cut put on. Before much grinding has taken place it is advisable to inspect the surface to see how it is cleaning up, as if the previous machining was reasonably correct the grinding should soon cover it. If, for example, the grinding is only occurring locally at one end or edge, the setting should be checked. When operating the vertical spindle cup wheel machine the side guards should always be in position, not only to catch the water, but also to protect the operator from injury should a component be flung off the magnetic chuck by the wheel. This risk is greater when small, thin articles are being ground.

**Examples of surface grinding**

*Example* 1. *To grind up a pair of parallel strips* 50 *mm* × 20 *mm* × 150 *mm.*

This is a plain surface job and may be done on a disc or cup wheel machine, the latter being somewhat quicker. Presumably the strips will have been previously cleaned up by milling or shaping. If not they may still be

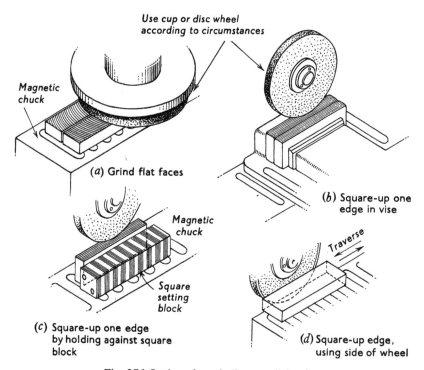

Fig. 276 Settings for grinding parallel strips

ground from the rough, but in such a case a vertical spindle cup wheel machine is recommended.

1. Inspect one face of the strips and remove any burrs. Place both strips with this face on a magnetic chuck and clean up the other face.

2. Turn the strips over and grind the opposite face to size, checking for flatness and parallelism (Fig. 276(a)).

3. One edge has now to be ground square with the faces just finished. This can be done with the strips held in a good vise as at (b), by making use of a square block on the magnetic chuck as at (c) or by working from the face of the chuck and the side of the wheel (d).

4. Hold on the magnetic chuck and finish the other edge as Fig. 276(a).

*Example 2. To grind up a pair of vee blocks to the sketch at Fig. 277(a).*

In this, as in many machining jobs, there are various ways of achieving the result and the method selected must depend upon the equipment available. In any case, the sequence and settings should be planned to yield

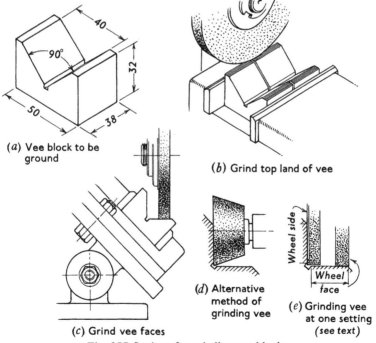

(a) Vee block to be ground

(b) Grind top land of vee

(c) Grind vee faces

(d) Alternative method of grinding vee

(e) Grinding vee at one setting (see text)

Fig. 277 Settings for grinding vee blocks

final accuracy in the important details with the minimum of effort and without recourse to chance. For this particular case none of the dimensions are important, but the heights of the vees on the two blocks must be exactly equal and the vee angles must be symmetrical about the centre line (i.e. they must not be 'drunken').

1. Grind up the side faces of the blocks parallel and equal. This can be done on a magnetic chuck.

2. Hold both blocks in a vise with the vee uppermost and grind the top land of the vee (Fig. 277(b)). Do not remove blocks from vise.

3. Set an adjustable angle plate at 45° using sine bar or vernier protractor, clamp it to the machine table and set it parallel with the travel, working from the table edge or tee slot.

4. Clamp the vise to the angle plate and set its length square with the edge of the plate. Grind one face of the vees (Fig. 277(c)). This could also be done with a cup wheel as at (d).

Check for parallelism between the vees and the side of the blocks by placing a test bar in the vees and measuring to the side, or to the face of the vise jaw.

5. Turn the vise through 180° and grind the other face of the vees.

(*Note.*—Both vee faces could be ground at one setting by grinding one face with the periphery and the other with the side of the wheel (Fig. 277(e)). Actually, this method offers best possibilities for accuracy, but results in a circular finish mark on one face and a straight one on the other.)

6. Remove the blocks from the vise, hold both together on the magnetic chuck and grind the base.

If it is necessary for the vee to be accurately centralised between the block sides this may now be adjusted. Set the blocks on one side, the same way round as when the vee was ground. Clamp a test bar into the vee and by measuring from the bar to the surface plate with slip gauges or height gauge, determine how much the vee is out of centre with the sides. Remove sufficient from the appropriate side to centralise the bar.

*Example* 3. *To grind the gauge shown at Fig. 278(a).*

1. Hold on tongue in vise and clean up the base. At the same time, after ensuring that the vise jaw is parallel with the table travel, clean up one edge (leave a witness if the base width has been shaped to size) (Fig. 278(b)).

2. Hold on the magnetic chuck and set the ground edge of the base parallel with the travel. Grind base to thickness, and with the sides of the

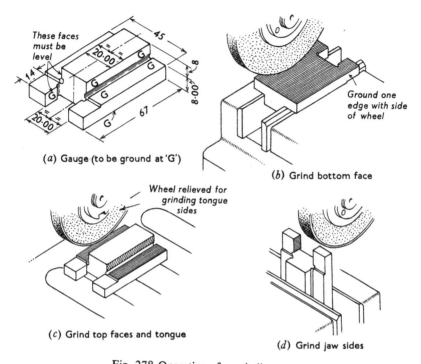

(a) Gauge (to be ground at 'G')

(b) Grind bottom face

(c) Grind top faces and tongue

(d) Grind jaw sides

Fig. 278 Operations for grinding gauge

wheel grind the tongue (Fig. 278(c)). Measure carefully the distance from the ground edge of the base to the nearer tongue side.

3. Hold vertically in the vise and set the tongue square with the table. Grind the side of the gap nearest the ground edge of the base until the measurement to the edge is the same as that from the edge to the tongue (Fig. 278(d)).

4. Finish the gap to width, checking with slip gauges.

## The tool and cutter grinder

This is a machine designed primarily for sharpening and reconditioning milling cutters, drills, reamers, taps and various other types of tools used in the shop. A more recent design of this machine, the Universal Tool and Cutter Grinder, is capable of undertaking a good range of surface grinding as well as the grinding of cutters, and with the necessary attachments may also be used for light external and internal grinding. It is possible to set up for grinding cutters on the universal cylindrical machine, but this machine is hardly ever used for such a purpose, as the tool and cutter grinder may be adapted to the settings with greater facility as well as incorporating in its design the lighter movements and quick wheel-changing arrangements necessary for the work. Furthermore, the treatment of tools and cutters requires a variety of special accessories which are readily available for the tool and cutter grinder.

A diagram of a universal tool and cutter grinder is shown at Fig. 279. The table has a separate swivelling top similar to that we have previously described on the universal machine and is hand-operated on a longitudinal slide. The table slides are carried on another set of cross-slides for giving

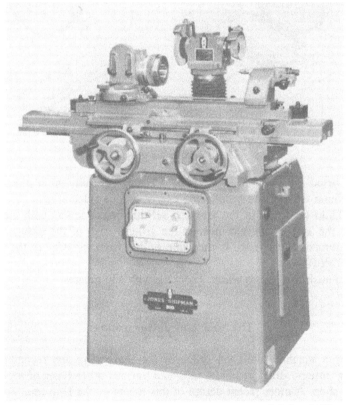

*A. A. Jones & Shipman Ltd*

Fig. 279 Universal tool and cutter grinding machine

the necessary cross movement towards the wheel-head and the handwheels at the front of the machine are for operating the longitudinal and cross movements. These two controls are duplicated at the rear of the machine since for many cutter grinding jobs the operator works from the rear of the machine and not from the front. For attaching the necessary fixtures the table-top is provided with a central tee slot. The wheel-head is mounted on a column behind the table, and the spindle is driven by a motor placed near the floor inside the machine body. The head and spindle may be swivelled round so as to bring either wheel over the table or at any intermediate angle. Vertical movement of the wheel-head is given by the handwheel placed on the left hand side of the machine. Equipment for use with the machine includes poppet centres for holding centred work or mandrils, vise with swivel base and angle bracket, a swivelling head with index plate for holding cutters and obtaining the tooth positions, a motor-driven head for cylindrical and internal grinding and various rests for supporting and locating the teeth of cutters being ground.

### Cutter grinding and setting

When a fluted tooth is being ground it is supported on a rest whilst the wheel passes over its edge. The arrangement may be as at Fig. 280(a) with the wheel rotating off the cutting edge, or as at (b) with the wheel meeting the edge. The advantage of the first method is that the wheel tends to hold the tooth down on the rest and being thus safer is more commonly used than (b), which gives a keener edge, free from burrs, and has less tendency to burn the edge. When method (a) is used the burrs should be

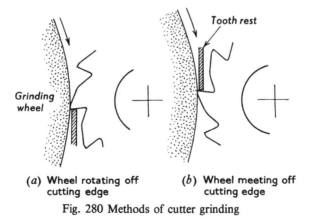

(a) Wheel rotating off       (b) Wheel meeting off
    cutting edge                 cutting edge

Fig. 280 Methods of cutter grinding

oilstoned off the tooth edges after grinding, whilst with method (b), care should be taken to hold the tooth against the rest.

## Clearance

The clearance angle is the angle $\alpha$ to which the narrow land of the cutting edge is ground (Fig. 281(a)). This land should be about 1 millimetre wide

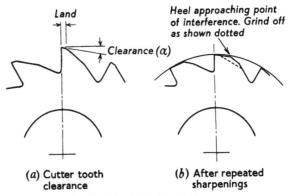

(a) Cutter tooth clearance      (b) After repeated sharpenings

Fig. 281 Tooth land and clearance

and when, with repeated grindings, it becomes too wide, a secondary clearance must be ground as shown at (b) to avoid the heel of the land fouling the work and interfering with cutting. The most suitable clearance angle depends mainly on the type and diameter of the cutter, and on the material it is to cut. A guide for clearance angles is as follows:

(a) *Cutter.* Up to 75 mm diameter: Clearance 6°–7°
     Over    „     „      „    4°–5°
     End teeth of end mills:   „    3°–5°

(b)

| Material being cut | Clearance |
|---|---|
| Low carbon steel . . . . . | 5°–7° |
| High carbon and alloy steel . . | 3°–5° |
| Cast iron and bronze. . . . | 4°–7° |
| Brass and aluminium . . . . | 10°–12° |

Insufficient clearance may result in the heel at the back of the land dragging against the work, whilst excessive clearance leads to rapid tooth wear, and promotes chatter.

**Setting for grinding**

The relative cutter-wheel position will depend upon whether grinding is being carried out with the periphery of a disc wheel or the face of a cup wheel.

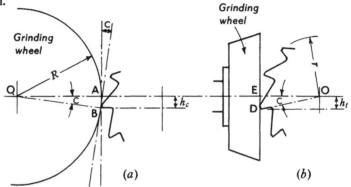

Fig. 282 Wheel and cutter setting for sharpening fluted cutter teeth

(a) *Disc wheel* (Fig. 282(a)).

In this case the rest (and tooth front) is set level with the cutter centre, but this is set below the wheel centre.

The clearance angle is shown as $C$, and $\dfrac{AB}{OB} = \sin C$

$\quad OB =$ wheel radius $(R)$

$\quad AB =$ offset $(h_c)$

Hence $\dfrac{h_c}{R} = \sin C$

$\quad h_c = R \sin C$

The procedure for setting-up is as follows:

1. Set the wheel and cutter centres level.

2. Clamp the tooth rest to the machine table and set it level with the cutter centre with a height gauge or special setting gauge provided.

3. Adjust the wheel or work up or down by the correct amount.

(b) *Cup wheel* (Fig. 282(b))

In triangle OED

$ED = h_c$;  $OD =$ cutter rad $(r)$

$\dfrac{ED}{OD} = \sin C$ or $\dfrac{h_c}{r} = \sin C$

$\quad h_t = r \sin C$

1. Adjust the cutter about central with the wheel, clamp the tooth rest to the wheel-head and set it level with the cutter centre.

2. Raise or lower the wheel-head and tooth rest by the correct amount.

A cup wheel produces a more definable result in all cases except where the land is narrow, as on wide lands the radiusing effect of the disc wheel, whilst giving the *average* clearance calculated, puts a falsely large clearance immediately behind the cutting edge (Fig. 283).

On cutters with their teeth cut to a fairly sharp helix the effect of this is to modify, in cutting, the clearance angle ground on the tooth; and the mathematical consideration of this is discussed on p. 118 of the author's *Senior Workshop Calculations.*

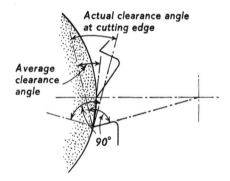

Fig. 283 Showing false clearance effect when
grinding a wide land with disc wheel

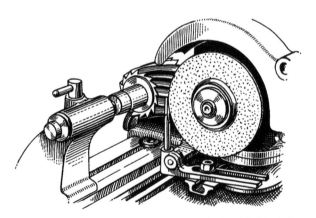

*The Norton Grinding Wheel Co, Ltd*

Fig. 284 Setting for sharpening a helical cylindrical cutter

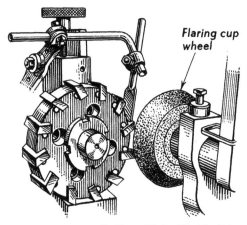

Flaring cup wheel

*The Norton Grinding Wheel Co, Ltd*

Fig. 285(a) Grinding blades on a face mill
(Similar arrangement applies to sides of S and F cutter teeth.)

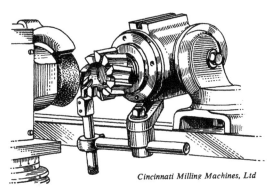

*Cincinnati Milling Machines, Ltd*

Fig. 285(b) Grinding end teeth on a shell end mill

## Set-ups for fluted type cutters

The arrangement for sharpening a cylindrical cutter is shown at Fig. 284. The cutter is held on a mandril between centres, and after the settings have been obtained one hand is used to hold the cutter and keep the tooth front in contact with the rest whilst the other hand operates the table traverse. As the tooth passes the wheel-face the rest rotates the cutter so that the twist of the tooth is followed. Two or three times round the cutter with about 0·05 mm cut each time should restore its edge. For grinding the side teeth of side and face mills and the end teeth of end mills a small cup wheel is most convenient, the cutter axis being set to the clearance angle. These are shown at Fig. 285.

### Reamers and taps

If a reamer is of the adjustable, expanding type it may be sharpened on the front of its teeth and then set out to its original size. Solid reamers, however, cannot be treated in this way as they would lose their size and must be sharpened by grinding the front faces of the teeth. To allow for

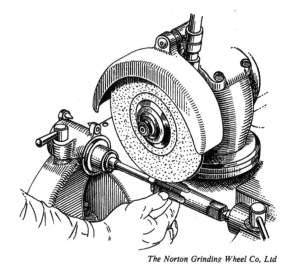

The Norton Grinding Wheel Co, Ltd

Fig. 286 Sharpening a tap (tooth rest just visible near thumb of operator)

this the lands are made wide enough to stand a reasonable amount being ground off before the clearance is reached. Taps must also be sharpened by grinding along their flutes with a formed wheel. Fig. 286 shows a tap being sharpened.

### Machine relieved cutters

When these cutters are made their tooth form is put on in a relieving lathe and when once obtained is permanent for the life of the cutter. No grinding must be done, therefore, which will affect the tooth profile, and the tooth is sharpened by grinding its front face. The cutting clearance is put on the tooth at the time of its formation and is unaffected by the sharpening. A narrow dish wheel is necessary to pass through the tooth gash, and when the tooth front is radial the face of the wheel must be set on the centre of the cutter before the tooth is brought up to it (Fig. 287(a)). When the tooth is gashed off centre for the purpose of providing front rake the wheel-face must be set over this amount, and the offset is sometimes marked on the cutter (Fig. 287(b)). To control the indexing for grinding the teeth a tooth-

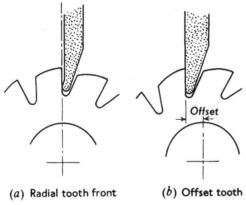

(a) Radial tooth front          (b) Offset tooth

Fig. 287 Sharpening machine relieved cutters

rest or a slotted indexing plate may be used. The indexing plate forms the most reliable method as it ensures that the tooth fronts are equally spaced, an important point upon which the concentricity of the cutter depends. If a tooth-rest is used it should be set against the *back* of a tooth (preferably the one being ground) and the tooth held against it. For this purpose, when a cutter is new, the backs of its teeth should be touched up by locating from the fronts so that a good supporting point is available for future grindings of the teeth.

When grinding one of these cutters the wheel-face should be set and a tooth face rotated until contact is made. The index wheel or tooth-rest is then set at this position and the cut taken across each tooth, traversing the cutter across the wheel-face with a steady hand operation of the table (Fig. 288). To add a little more cut for the next go round the cross-slide must not be moved, but the cutter given a slight rotation towards the wheel by advancing the tooth-rest. To compensate for wheel wear the last cut should be very light, a rule applying to all classes of cutter grinding. We mentioned above that the concentricity of the teeth depends on the accuracy of their spacing. As a check on this a small-testing fixture is useful and an example of such is shown at Fig. 289. The cutter is set on the peg

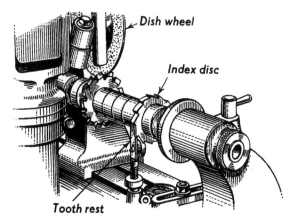

Fig. 288 Set-up for sharpening relieved cutter (index disc being used to space teeth)

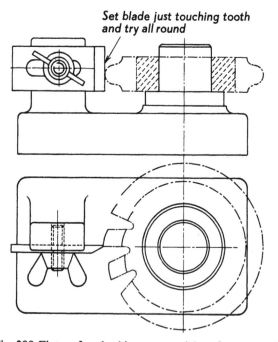

Fig. 289 Fixture for checking concentricity of cutter teeth

and the blade adjusted to the tooth edge, so that by trying each tooth in turn their trueness may be checked. If one or a number of teeth are found to be high a small additional amount must be ground off their fronts.

## Cutting-off

The perfection of plastics, shellac and rubber bonded wheels has brought the process of cutting-off within the sphere of the abrasive wheel and it is now being used extensively for cutting solid metal and tubing up to 75 mm diameter. The process is particularly suited to such materials as high carbon and alloy steel bars, chilled iron, thin tubing, etc., which have always been severe on saw teeth. In addition to cutting engineering materials, abrasive wheels are being employed in other industries for cutting such materials as asbestos, brick, china, glass, marble and so on.

Cutting-off machines may be classified as (a) high-speed machines operating the cutting wheels up to a peripheral speed of 5000 metres per min, (b) low-speed machines using speeds from 2750 m to 3750 m per min, and (c) portable machines used in stone yards, quarries, etc., and some- times in engineering shops for cutting light bars and tubing, snagging castings, etc. The process may be conducted wet or dry. Wet grinding helps to minimise the extreme heat generated and reduces the subsequent dis- coloration of the cut surface, but as the cut being made is filled by the abrasive wheel it is not possible to attain good conditions of cooling such as in the case of cylindrical grinding, and some discoloration is nearly always caused.

Plastics bonded wheels are most commonly used for the general run of work on these machines as this type of wheel is found to give the greatest satisfaction. Shellac wheels are used in the smaller sizes for special tool- room jobs, such as for grinding the sides of narrow slots and opening out the sawcuts in spring collets where the end of the cut must be left solid to prevent hardening distortion and enable the diameters of the collet to be ground after hardening. Rubber bonded wheels are also used for the above purposes and also occasionally for special cutting-off jobs. Shellac and rubber wheels can be produced in softer and more free-cutting qualities than hard plastics and this permits them to operate with less discoloration, but being softer these wheels have a shorter life and this may render the cost of the process uneconomic.

In the general machine shop and toolroom this method of cutting-off is extremely useful for high carbon, high-speed and alloy steel bars, and lengths may be cut to close limits. For reclaiming broken drills, taps, tool-bits, etc., the process is very useful as the damaged portion may be cut off and the remainder of the tool ground up for further use. The enormous heat generated during the passage of the wheel through the work generally raises the sides of the cut to a red heat and the rapid cooling effect of heat conduction to the cold adjoining metal, aided by the coolant if this is used, results in an effect almost equivalent to that of heating and quenching for hardening. The process, therefore, may result in some hardening on the cut ends of self-hardening steels, a tendency more pronounced when water cooling is employed.

A very useful application of the process has been for the dressing of runners and risers from foundry castings where the metal to be cut is often composed of chilled iron or hard bronze, both being notoriously severe on saws. The machine shown at Fig. 290 is one which has been developed for

foundry work, although with the auxiliary table and vise it can be used as a general-purpose machine for bars and tubes. The wheel used is 400 mm diameter, 3 mm thick and operates at a peripheral speed of 4600 metres per min driven by vee belts from a 7·5 kW motor. It is a characteristic of these machines that when dealing with large sections of soft material the peak power requirement may rise to about three times the normal motor rating and this is taken care of by using a motor having special electrical characteristics.

The following approximate cutting times are claimed for the process: 35 mm high-speed steel, 4–5 sec; 50 mm square stainless steel, 8–10 sec; 75 mm mild steel tube, 5–6 sec. The makers also produce another type of cutting-off machine on which the wheel-feed motion and bar clamping are operated hydraulically. In this machine an inclined threaded rod carries stop collars for controlling the limits of the wheel-head movement.

### Oilstones

A discussion of abrasives would be incomplete without some mention of oilstones which have a wide range of use in the machine shop. These are made in a large variety of sizes and shapes, generally with alumina abrasive, vitrified process, in fine, medium and coarse grits. They are used extensively by toolmakers for fine finishing work on gauges, tools, punches, dies, etc., but the tool kit of any mechanic is incomplete without the inclusion of a small selection. An ordinary cutting tool after grinding may have a slight burr thrown up on the cutting edge. By removing this and smoothing the edge with a stone the tool will cut better, give a smoother finish and stand up longer to its work. Milling cutters, reamers, etc., are improved after grinding by oilstoning the front face of the cutting edge. The many other general uses to which oilstones may be put are too numerous to mention: cutlery is kept in condition by an occasional application, high spots may be removed from shafts when fitting new bearings, and so on. As the name implies, oil is used as a lubricant with these stones, and if very free cutting qualities are desired paraffin may be used.

## Grinding wheel recommendations

The following brief list is added as a guide to the selection of grinding wheels. The gradings given are to the Norton recommendation and are specified according to the British Standard method shown on p. 316, with prefix and suffix (where given) particular to this maker.

| TYPE OF GRINDING AND METAL | WHEEL |
|---|---|

*Cylindrical grinding*

| Hard steel . . . . . . . . | A60–L5VG |
| Soft steel . . . . . . . . | A60–N5VG |
| Cast iron . . . . . . . . | 37C54–LVK |
| Aluminium ⎫ Brass and bronze ⎬ . . . . . | 37C46–KVK |

*Surface grinding*

Plain wheels used on periphery

| Hard steel . . . . . . . . | 38A46–H8VBE |
| Soft steel . . . . . . . . | 38A46–J5VBE |
| General purpose and Cast iron ⎬ . . . . . | 38A46–I8VBE |
| Aluminium Brass and bronze ⎬ . . . . . | 38A46–F12VBEP |

Cup and cylinder wheels

| Hard steel . . . . . . . . | 38A30–H8VBE |
| Soft steel . . . . . . . . | 38A30–J8VBE |
| General purpose and Cast iron ⎬ . . . . . | 38A30–I8VBE |

*Internal grinding*

| Hard steel . . . . . . . . | 32A60–K5VG |
| Soft steel . . . . . . . . | 19A60–M5VBE |
| Cast iron . . . . . . . . | 32A60–L5VG |
| Aluminium Brass and bronze . . . . . | 32A54–L5VBE |

*Cutter grinding*

| Milling cutters, reamers, etc. (HSS). . . | 38A46–K5VG |
| Cemented carbides | |
| Roughing . . . . . . . . | 39C60–J8VK |
| Finishing . . . . . . . . | 39C100–J8VK |
| Diamond wheel (general use). . . . | D150–N50B |

*Drill grinding*
    Large drills . . . . . . . . . A46–M5VBE
    Small drills . . . . . . . . . A80–L5VBE

*Taps* (sharpening)
    Large . . . . . . . . . 38A46–N7VG
    Small . . . . . . . . . 38A80–N7VG

*Off-hand grinding*
  Bench and pedestal grinders
    Roughing (general purpose) . . . . A36–P5VBE
    Finishing . . . . . . . . . A60–N5VBE
  Tool grinders
    300 to 600 mm diameter wheels
    Wet grinding . . . . . . . A30–05VBE
*Cutting-Off*—at 3000 to 5000 surface metres per minute
  General purpose
  250 to 400 mm diameter
    Dry . . . . . . . . . . A30–T6B9FS
    Wet . . . . . . . . . . A46–Q8R50

## Conclusion

We hope that the reader, by the time he has mastered the practical technique connected with this and our first volume, will have realised that the work of the machine shop demands high qualities in thought, integrity and skill from those engaged on it. The road to the attainment of the highest practical and theoretical reaches in the making of an engineer is hard and long, but no one who is worth his salt would have it otherwise, as it is only on such foundations that a lasting and satisfactory social structure may be built up. Let us urge the reader, therefore, to avoid all half-measures in his activities. When at work let it be at full efficiency. When the time comes for recreation let that also be undertaken in the spirit befitting to the occasion, whether it be quiet rest or active games. By so ordering his life the reader should live to look back on his efforts with a degree of satisfaction, and whether he attains greatness or remains a simple worker will not matter. To him, happiness will come anyway, and when the time arrives to give up he will be able to do so in the knowledge that the community has benefited by his efforts.

# Conversion tables

## Fractional Sub-divisions of an inch to decimals and to millimetres.

| in | in | milli-metres | in | in | milli-metres |
|---|---|---|---|---|---|
| 1/64 | 0·015625 | 0·3969 | 39/64 | 0·609375 | 15·4781 |
| 1/32 | 0·03125 | 0·7938 | 5/8 | 0·625 | 15·875 |
| 3/64 | 0·046875 | 1·1906 | 41/64 | 0·640625 | 16·2719 |
| 1/16 | 0·0625 | 1·5875 | 21/32 | 0·65625 | 16·6688 |
| 5/64 | 0·078125 | 1·9844 | 43/64 | 0·671875 | 17·0656 |
| 3/32 | 0·09375 | 2·3812 | 11/16 | 0·6875 | 17·4625 |
| 7/64 | 0·109375 | 2·7781 | 45/64 | 0·703125 | 17·8594 |
| 1/8 | 0·125 | 3·175 | 23/32 | 0·71875 | 18·2562 |
| 9/64 | 0·140625 | 3·5719 | 47/64 | 0·734375 | 18·6531 |
| 5/32 | 0·15625 | 3·9688 | 3/4 | 0·75 | 19·05 |
| 11/64 | 0·171875 | 4·3656 | 49/64 | 0·765625 | 19·4469 |
| 3/16 | 0·1875 | 4·7625 | 25/32 | 0·78125 | 19·8438 |
| 13/64 | 0·203125 | 5·1594 | 51/64 | 0·796875 | 20·2406 |
| 7/32 | 0·21875 | 5·5562 | 13/16 | 0·8125 | 20·6375 |
| 15/64 | 0·234375 | 5·9531 | 53/64 | 0·828125 | 21·0344 |

## Millimetres to inches. Based on 1 inch = 25·4 millimetres

| mm | 0 | 1 | 2 | 3 | 4 | 5 | 6 | 7 | 8 | 9 |
|---|---|---|---|---|---|---|---|---|---|---|
| | in | in | in | in | in | in | in | in | in | in |
| – | – | 0·03937 | 0·07874 | 0·11811 | 0·15748 | 0·19685 | 0·23622 | 0·27559 | 0·31496 | 0·35433 |
| 10 | 0·39370 | 0·43307 | 0·47244 | 0·51181 | 0·55118 | 0·59055 | 0·62992 | 0·66929 | 0·70866 | 0·74803 |
| 20 | 0·78740 | 0·82677 | 0·86614 | 0·90551 | 0·94488 | 0·98425 | 1·02362 | 1·06299 | 1·10236 | 1·14173 |
| 30 | 1·18110 | 1·22047 | 1·25984 | 1·29921 | 1·33858 | 1·37795 | 1·41732 | 1·45669 | 1·49606 | 1·53543 |
| 40 | 1·57480 | 1·61417 | 1·65354 | 1·69291 | 1·73228 | 1·77165 | 1·81102 | 1·85039 | 1·88976 | 1·92913 |
| 50 | 1·96850 | 2·00787 | 2·04724 | 2·08661 | 2·12598 | 2·16535 | 2·20472 | 2·24409 | 2·28346 | 2·32283 |
| 60 | 2·36220 | 2·40157 | 2·44094 | 2·48031 | 2·51969 | 2·55906 | 2·59843 | 2·63780 | 2·67717 | 2·71654 |
| 70 | 2·75591 | 2·79528 | 2·83465 | 2·87402 | 2·91339 | 2·95276 | 2·99213 | 3·03150 | 3·07087 | 3·11024 |
| 80 | 3·14961 | 3·18898 | 3·22835 | 3·26772 | 3·30709 | 3·34646 | 3·38583 | 3·42520 | 3·46457 | 3·50394 |
| 90 | 3·54331 | 3·58268 | 3·62205 | 3·66142 | 3·70079 | 3·74016 | 3·77953 | 3·81890 | 3·85827 | 3·89764 |
| 100 | 3·93701 | 3·97638 | 4·01575 | 4·05512 | 4·09449 | 4·13386 | 4·17323 | 4·21260 | 4·25197 | 4·29134 |
| 10 | 4·33071 | 4·37008 | 4·40945 | 4·44882 | 4·48819 | 4·52756 | 4·56693 | 4·60630 | 4·64567 | 4·68504 |
| 20 | 4·72441 | 4·76378 | 4·80315 | 4·84252 | 4·88189 | 4·92126 | 4·96063 | 5·0000 | 5·0394 | 5·0787 |
| 30 | 5·1181 | 5·1575 | 5·1969 | 5·2362 | 5·2756 | 5·3150 | 5·3543 | 5·3937 | 5·4331 | 5·4724 |
| 40 | 5·5118 | 5·5512 | 5·5906 | 5·6299 | 5·6693 | 5·7087 | 5·7480 | 5·7874 | 5·8268 | 5·8661 |
| 50 | 5·9055 | 5·9449 | 5·9843 | 6·0236 | 6·0630 | 6·1024 | 6·1417 | 6·1811 | 6·2205 | 6·2598 |
| 60 | 6·2992 | 6·3386 | 6·3780 | 6·4173 | 6·4567 | 6·4961 | 6·5354 | 6·5748 | 6·6142 | 6·6535 |
| 70 | 6·6929 | 6·7323 | 6·7717 | 6·8110 | 6·8504 | 6·8898 | 6·9291 | 6·9685 | 7·0079 | 7·0472 |
| 80 | 7·0866 | 7·1260 | 7·1654 | 7·2047 | 7·2441 | 7·2835 | 7·3228 | 7·3622 | 7·4016 | 7·4409 |
| 90 | 7·4803 | 7·5197 | 7·5591 | 7·5984 | 7·6378 | 7·6772 | 7·7165 | 7·7559 | 7·7953 | 7·8346 |

## Left conversion table

| Fraction | in | mm | Fraction | in | mm |
|---|---|---|---|---|---|
| 1/4 | 0·25 | 6·35 | 27/32 | 0·84375 | 21·4312 |
| 17/64 | 0·265625 | 6·7469 | 55/64 | 0·859375 | 21·8281 |
| 9/32 | 0·28125 | 7·1438 | 7/8 | 0·875 | 22·225 |
| 19/64 | 0·296875 | 7·5406 | 57/64 | 0·890625 | 22·6219 |
| 5/16 | 0·3125 | 7·9375 | 29/32 | 0·90625 | 23·0188 |
| 21/64 | 0·328125 | 8·3344 | 59/64 | 0·921875 | 23·4156 |
| 11/32 | 0·34375 | 8·7312 | 15/16 | 0·9375 | 23·8125 |
| 23/64 | 0·359375 | 9·1281 | 61/64 | 0·953125 | 24·2094 |
| 3/8 | 0·375 | 9·525 | 31/32 | 0·96875 | 24·6062 |
| 25/64 | 0·390625 | 9·9219 | 63/64 | 0·984375 | 25·0031 |
| 13/32 | 0·40625 | 10·3188 | 1 | 1 | 25·4 |
| 27/64 | 0·421875 | 10·7156 | 2 | 2 | 50·800 |
| 7/16 | 0·4375 | 11·1125 | 3 | 3 | 76·200 |
| 29/64 | 0·453125 | 11·5094 | 4 | 4 | 101·600 |
| 15/32 | 0·46875 | 11·9062 | 5 | 5 | 127·000 |
| 31/64 | 0·484375 | 12·3031 | 6 | 6 | 152·400 |
| 1/2 | 0·5 | 12·7 | 7 | 7 | 177·800 |
| 33/64 | 0·515625 | 13·0969 | 8 | 8 | 203·200 |
| 17/32 | 0·53125 | 13·4938 | 9 | 9 | 228·600 |
| 35/64 | 0·546875 | 13·8906 | 10 | 10 | 254·000 |
| 9/16 | 0·5625 | 14·2875 | 11 | 11 | 279·400 |
| 37/64 | 0·578125 | 14·6844 | 12 | 12 | 304·800 |
| 19/32 | 0·59375 | 15·0812 | | | |

## Right conversion table

| 200 | 7·8740 | 7·9134 | 7·9528 | 7·9921 | 8·0315 | 8·0709 | 8·1102 | 8·1496 | 8·1890 | 8·2283 |
|---|---|---|---|---|---|---|---|---|---|---|
| 10 | 8·2677 | 8·3071 | 8·3465 | 8·3858 | 8·4252 | 8·4646 | 8·5039 | 8·5433 | 8·5827 | 8·6220 |
| 20 | 8·6614 | 8·7008 | 8·7402 | 8·7795 | 8·8189 | 8·8583 | 8·8976 | 8·9370 | 8·9764 | 9·0157 |
| 30 | 9·0551 | 9·0945 | 9·1339 | 9·1732 | 9·2126 | 9·2520 | 9·2913 | 9·3307 | 9·3701 | 9·4094 |
| 40 | 9·4488 | 9·4882 | 9·5276 | 9·5669 | 9·6063 | 9·6457 | 9·6850 | 9·7244 | 9·7638 | 9·8031 |
| 50 | 9·8425 | 9·8819 | 9·9213 | 9·9606 | 10·0000 | 10·0394 | 10·0787 | 10·1181 | 10·1575 | 10·1969 |
| 60 | 10·2362 | 10·2756 | 10·3150 | 10·3543 | 10·3937 | 10·4331 | 10·4724 | 10·5118 | 10·5512 | 10·5906 |
| 70 | 10·6299 | 10·6693 | 10·7087 | 10·7480 | 10·7874 | 10·8268 | 10·8661 | 10·9055 | 10·9449 | 10·9843 |
| 80 | 11·0236 | 11·0630 | 11·1024 | 11·1417 | 11·1811 | 11·2205 | 11·2598 | 11·2992 | 11·3386 | 11·3780 |
| 90 | 11·4173 | 11·4567 | 11·4961 | 11·5354 | 11·5748 | 11·6142 | 11·6535 | 11·6929 | 11·7323 | 11·7717 |
| 300 | 11·8110 | 11·8504 | 11·8898 | 11·9291 | 11·9685 | 12·0079 | 12·0472 | 12·0866 | 12·1260 | 12·1654 |
| 10 | 12·2047 | 12·2441 | 12·2835 | 12·3228 | 12·3622 | 12·4016 | 12·4409 | 12·4803 | 12·5197 | 12·5591 |
| 20 | 12·5984 | 12·6378 | 12·6772 | 12·7165 | 12·7559 | 12·7953 | 12·8346 | 12·8740 | 12·9134 | 12·9528 |
| 30 | 12·9921 | 13·0315 | 13·0709 | 13·1102 | 13·1496 | 13·1890 | 13·2283 | 13·2677 | 13·3071 | 13·3465 |
| 40 | 13·3858 | 13·4252 | 13·4646 | 13·5039 | 13·5433 | 13·5827 | 13·6220 | 13·6614 | 13·7008 | 13·7402 |
| 50 | 13·7795 | 13·8189 | 13·8583 | 13·8976 | 13·9370 | 13·9764 | 14·0157 | 14·0551 | 14·0945 | 14·1339 |
| 60 | 14·1732 | 14·2126 | 14·2520 | 14·2913 | 14·3307 | 14·3701 | 14·4094 | 14·4488 | 14·4882 | 14·5276 |
| 70 | 14·5669 | 14·6063 | 14·6457 | 14·6850 | 14·7244 | 14·7638 | 14·8031 | 14·8425 | 14·8819 | 14·9213 |
| 80 | 14·9606 | 15·0000 | 15·0394 | 15·0787 | 15·1181 | 15·1575 | 15·1969 | 15·2362 | 15·2756 | 15·3150 |
| 90 | 15·3543 | 15·3937 | 15·4331 | 15·4724 | 15·5118 | 15·5512 | 15·5906 | 15·6299 | 15·6693 | 15·7087 |
| 400 | 15·7480 | 15·7874 | 15·8268 | 15·8661 | 15·9055 | 15·9449 | 15·9843 | 16·0236 | 16·0630 | 16·1024 |
| 10 | 16·1417 | 16·1811 | 16·2205 | 16·2598 | 16·2992 | 16·3386 | 16·3780 | 16·4173 | 16·4567 | 16·4961 |
| 20 | 16·5354 | 16·5748 | 16·6142 | 16·6535 | 16·6929 | 16·7323 | 16·7717 | 16·8110 | 16·8504 | 16·8898 |
| 30 | 16·9291 | 16·9685 | 17·0079 | 17·0472 | 17·0866 | 17·1260 | 17·1654 | 17·2047 | 17·2441 | 17·2835 |
| 40 | 17·3228 | 17·3622 | 17·4016 | 17·4409 | 17·4803 | 17·5197 | 17·5591 | 17·5984 | 17·6378 | 17·6772 |
| 50 | 17·7165 | 17·7559 | 17·7953 | 17·8346 | 17·8740 | 17·9134 | 17·9528 | 17·9921 | 18·0315 | 18·0709 |
| 60 | 18·1102 | 18·1496 | 18·1890 | 18·2283 | 18·2677 | 18·3071 | 18·3465 | 18·3858 | 18·4252 | 18·4646 |
| 70 | 18·5039 | 18·5433 | 18·5827 | 18·6220 | 18·6614 | 18·7008 | 18·7402 | 18·7795 | 18·8188 | 18·8583 |
| 80 | 18·8976 | 18·9370 | 18·9764 | 19·0157 | 19·0551 | 19·0945 | 19·1339 | 19·1732 | 19·2126 | 19·2520 |
| 90 | 19·2913 | 19·3307 | 19·3701 | 19·4094 | 19·4488 | 19·4882 | 19·5276 | 19·5669 | 19·6063 | 19·6457 |
| 500 | 19·6850 | 19·7244 | 19·7638 | 19·8031 | 19·8425 | 19·8819 | 19·9213 | 19·9606 | 20·0000 | 20·0394 |

GKN Bolts & Nuts Ltd.

For a full list of conversions refer to BS 350, Parts 1 and 2.

## APPENDIX 2

### Brown and Sharpe Tapers

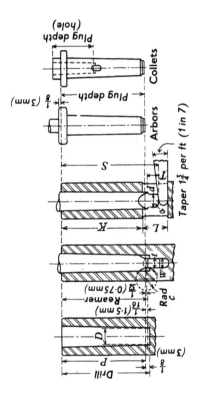

| No of taper | Plug dia small end D (mm) | Plug dia small end D (in) | Plug depth P† (mm) | Plug depth P† (in) | Keyway from end of spindle K (mm) | Keyway from end of spindle K (in) | Shank depth S (mm) | Shank depth S (in) | Length of keyway L* (mm) | Length of keyway L* (in) | Width of keyway W (mm) | Width of keyway W (in) | Length of tongue T (mm) | Length of tongue T (in) | Dia of tongue d (mm) | Dia of tongue d (in) | Thickness of tongue t (mm) | Thickness of tongue t (in) | Radius of tongue cutter c (mm) | Radius of tongue cutter c (in) | Radius of tongue end a (mm) | Radius of tongue end a (in) |
|---|---|---|---|---|---|---|---|---|---|---|---|---|---|---|---|---|---|---|---|---|---|---|
| 1 | 5·08 | ·200 | 23·8 | 15/16 | 23·8 | 15/16 | 30 | 1 3/16 | 9·5 | 3/8 | 3·4 | ·135 | 4·8 | 3/16 | 4·3 | ·170 | 3·2 | 1/8 | 4·8 | 3/16 | ·75 | ·03 |
| 2 | 6·35 | ·250 | 30·2 | 1 3/16 | 29·8 | 1 3/16 | 38 | 1 1/2 | 12·7 | 1/2 | 4·2 | ·166 | 6·4 | 1/4 | 5·6 | ·220 | 4 | 5/32 | 4·8 | 3/16 | ·75 | ·03 |
| 3 | 7·94 | ·312 | 38 | 1 1/2 | 37 | 1 7/16 | 47·5 | 1 7/8 | 16 | 5/8 | 5 | ·197 | 8 | 5/16 | 7·2 | ·282 | 4·8 | 3/16 | 4·8 | 3/16 | 1 | ·04 |
| 4 | 8·89 | ·350 | 43 | 1 11/16 | 41·6 | 1 5/8 | 53 | 2 3/32 | 17·5 | 11/16 | 5·8 | ·228 | 8·7 | 11/32 | 8·1 | ·320 | 5·5 | 7/32 | 8 | 5/16 | 1·25 | ·05 |
| 5 | 11·43 | ·450 | 53 | 2 1/8 | 52·5 | 2 1/16 | 65 | 2 9/16 | 19 | 3/4 | 6·6 | ·260 | 9·5 | 3/8 | 10·7 | ·420 | 6·4 | 1/4 | 8 | 5/16 | 1·5 | ·06 |
| 6 | 12·70 | ·500 | 60 | 2 3/8 | 58·5 | 2 5/16 | 73 | 2 7/8 | 22 | 7/8 | 7·4 | ·291 | 11·1 | 7/16 | 11·7 | ·460 | 7·1 | 9/32 | 8 | 5/16 | 1·5 | ·06 |
| 7 | 15·24 | ·600 | 73 | 2 7/8 | 71 | 2 25/32 | 86·5 | 3 13/32 | 24 | 15/16 | 8·2 | ·322 | 11·9 | 15/32 | 14·2 | ·560 | 8· | 5/16 | 9·5 | 3/8 | 1·75 | ·07 |
| 8 | 19·05 | ·750 | 90·5 | 3 9/16 | 87·5 | 3 7/16 | 105 | 4 1/8 | 25·5 | 1 | 9 | ·353 | 12·7 | 1/2 | 18 | ·710 | 8·7 | 11/32 | 9·5 | 3/8 | 2 | ·08 |
| 9 | 22·86 | ·900 | 108 | 4 1/4 | 105 | 4 1/8 | 124 | 4 7/8 | 28·5 | 1 1/8 | 9·8 | ·385 | 14·3 | 9/16 | 21·8 | ·860 | 9·5 | 3/8 | 11·1 | 7/16 | 2·5 | ·10 |
| 10 | 26·54 | 1·045 | 127 | 5 | 123 | 4 27/32 | 145 | 5 23/32 | 33·3 | 1 5/16 | 11·3 | ·447 | 16·7 | 21/32 | 25·6 | 1·010 | 11·1 | 7/16 | 11·1 | 7/16 | 2·8 | ·11 |
| 11 | 31·75 | 1·250 | 150 | 5 15/16 | 147 | 5 25/32 | 169 | 6 21/32 | 33·3 | 1 5/16 | 11·3 | ·447 | 16·7 | 21/32 | 30·8 | 1·210 | 11·1 | 7/16 | 12·7 | 1/2 | 3·3 | ·13 |
| 12 | 38·10 | 1·500 | 181 | 7 1/8 | 176 | 6 15/16 | 201 | 7 15/16 | 38 | 1 1/2 | 13 | ·510 | 19 | 3/4 | 37 | 1·460 | 12·7 | 1/2 | 12·7 | 1/2 | 3·8 | ·15 |
| 13 | 44·45 | 1·750 | 197 | 7 3/4 | 192 | 7 9/16 | 218 | 8 9/16 | 38 | 1 1/2 | 13 | ·510 | 19 | 3/4 | 43·5 | 1·710 | 12·7 | 1/2 | 16 | 5/8 | 4·3 | ·17 |
| 14 | 50·80 | 2·00 | 210 | 8 1/4 | 204 | 8 1/16 | 232 | 9 5/32 | 43 | 1 11/16 | 14·5 | ·572 | 21·4 | 27/32 | 49·8 | 1·960 | 14·3 | 9/16 | 19 | 3/4 | 4·8 | ·19 |
| 15 | 57·15 | 2·250 | 222 | 8 3/4 | 216 | 8 1/2 | 245 | 9 21/32 | 43 | 1 11/16 | 14·5 | ·572 | 21·4 | 27/32 | 56·2 | 2·210 | 14·3 | 9/16 | 22·2 | 7/8 | 5·3 | ·21 |
| 16 | 63·50 | 2·500 | 235 | 9 1/4 | 228 | 9 | 261 | 10 1/4 | 47·5 | 1 7/8 | 16 | ·635 | 23·8 | 15/16 | 62·3 | 2·450 | 16 | 5/8 | 25·4 | 1 | 5·8 | ·23 |

* Special lengths used in some cases. Standard lengths need not be used when keyway is for driving only, and not when using drift for extracting shank.
† Standard plug depths not used in all cases.

# APPENDIX 3

## ISO System of Limits and Fits. (BS 4500 : 1969)

Tolerance limits for selected holes. (Hole basis.)

| Nominal sizes | | H7 | | H8 | | H9 | | H11 | |
|---|---|---|---|---|---|---|---|---|---|
| Over mm | Up to and incl. mm | ul + | ll | ul + | ll | ul + | ll | ul + | ll |
| 6 | 10 | 15 | 0 | 22 | 0 | 36 | 0 | 90 | 0 |
| 10 | 18 | 18 | 0 | 27 | 0 | 43 | 0 | 110 | 0 |
| 18 | 30 | 21 | 0 | 33 | 0 | 52 | 0 | 130 | 0 |
| 30 | 50 | 25 | 0 | 39 | 0 | 62 | 0 | 160 | 0 |
| 50 | 80 | 30 | 0 | 46 | 0 | 74 | 0 | 190 | 0 |
| 80 | 120 | 35 | 0 | 54 | 0 | 87 | 0 | 220 | 0 |
| 120 | 180 | 40 | 0 | 63 | 0 | 100 | 0 | 250 | 0 |
| 180 | 250 | 46 | 0 | 72 | 0 | 115 | 0 | 290 | 0 |

ul = Upper limit; ll = Lower limit; Unit = 0·001 mm

Tolerance limits for selected shafts

| Nominal sizes | | c11 | | d10 | | e9 | | f7 | | g6 | | h6 | | k6 | | n6 | | p6 | | s6 | |
|---|---|---|---|---|---|---|---|---|---|---|---|---|---|---|---|---|---|---|---|---|---|
| Over mm | To mm | ul − | ll − | ul − | ll − | ul − | ll − | ul − | ll − | ul − | ll − | ul − | ll − | ul + | ll + | ul + | ll + | ul + | ll + | ul + | ll + |
| 6 | 10 | 80 | 170 | 40 | 98 | 25 | 61 | 13 | 28 | 5 | 14 | 0 | 9 | 10 | 1 | 19 | 10 | 24 | 15 | 32 | 23 |
| 10 | 18 | 95 | 205 | 50 | 120 | 32 | 75 | 16 | 34 | 6 | 17 | 0 | 11 | 12 | 1 | 23 | 12 | 29 | 18 | 39 | 28 |
| 18 | 30 | 110 | 240 | 65 | 149 | 40 | 92 | 20 | 41 | 7 | 20 | 0 | 13 | 15 | 2 | 28 | 15 | 35 | 22 | 48 | 35 |
| 30 | 40 | 120 | 280 | 80 | 180 | 50 | 112 | 25 | 50 | 9 | 25 | 0 | 16 | 18 | 2 | 33 | 17 | 42 | 26 | 59 | 43 |
| 40 | 50 | 130 | 290 | | | | | | | | | | | | | | | | | | |
| 50 | 65 | 140 | 330 | 100 | 220 | 60 | 134 | 30 | 60 | 10 | 29 | 0 | 19 | 21 | 2 | 39 | 20 | 51 | 32 | 72 | 53 |
| 65 | 80 | 150 | 340 | | | | | | | | | | | | | | | | | 78 | 59 |
| 80 | 100 | 170 | 390 | 120 | 260 | 72 | 159 | 36 | 71 | 12 | 34 | 0 | 22 | 25 | 3 | 45 | 23 | 59 | 37 | 93 | 71 |
| 100 | 120 | 180 | 400 | | | | | | | | | | | | | | | | | 101 | 79 |
| 120 | 140 | 200 | 450 | 145 | 305 | 85 | 185 | 43 | 83 | 14 | 39 | 0 | 25 | 28 | 3 | 52 | 27 | 68 | 43 | 117 | 92 |
| 140 | 160 | 210 | 460 | | | | | | | | | | | | | | | | | 125 | 100 |
| 160 | 180 | 230 | 480 | 170 | 355 | 100 | 215 | 50 | 96 | 15 | 44 | 0 | 29 | 33 | 4 | 60 | 31 | 79 | 50 | 133 | 108 |
| 180 | 200 | 240 | 530 | | | | | | | | | | | | | | | | | 151 | 122 |
| 200 | 225 | 260 | 550 | | | | | | | | | | | | | | | | | 159 | 130 |
| 225 | 250 | 280 | 570 | | | | | | | | | | | | | | | | | 169 | 140 |

ul = Upper limit, ll = Lower limit, Unit = 0·001 mm

# APPENDIX 4

## ISO System of Limits and Fits (BS 4500 : 1969)

Hole and shaft relationships for selected fits. (Hole basis.)

(Tolerance scale applies to the diameter range, 18 mm to 30 mm)

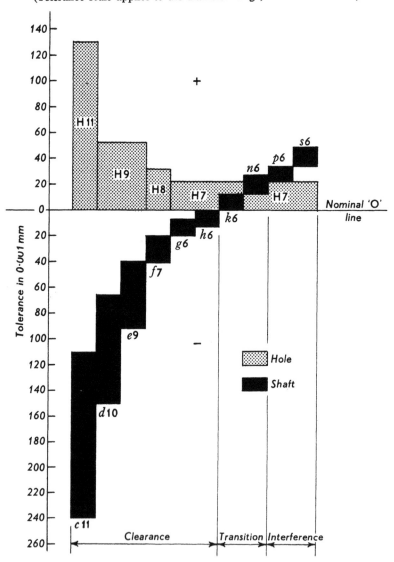

# APPENDIX 5

## List of British Standard Specifications of Interest to the Workshop Engineer

(Readers should consult the BSI handbook and index regularly as additions and amendments to specifications are constantly taking place.)

| BS No. | Publication Date | Title |
|---|---|---|
| 122 Pt. 1, | 1953 | Milling cutters |
| Pt. 2, | 1964 | Reamers, Countersinks and Counterbores |
| 308 | 1964 | Engineering Drawing Practice |
| 328 Pt. 1, | 1959 | Twist Drills |
| Pt. 2, | 1950 | Combined Drills and Countersinks (Centre Drills) |
| 426 | 1969 | Fixed Centres for use on Machine Tools Etc. |
| 498 Pt. 1, | 1960 | Files and Rasps |
| 620 | 1954 | Dimensions of Grinding Wheels and Segments of G.W. |
| 817 | 1957 | Cast Iron and Granite Surface Plates |
| 818 | 1963 | Cast Iron Straightedges |
| 852 | 1939 | Toolmakers' Straightedges |
| 863 | 1939 | Steel Straightedges of Rectangular Section |
| 869 | 1939 | Toolmakers' Flats and High Precision Surface Plates |
| 870 | 1950 | Micrometers (External) |
| 876 | 1964 | Hand Hammers |
| 887 | 1950 | Vernier Calipers |
| 888 | 1950 | Slip (or Block) Gauges and their Accessories |
| 906 | 1940 | Engineers' Parallels (Steel) |
| 907 | 1965 | Dial Gauges for Linear Measurements |
| 919 Pt. 1, | 1960 | Gauges for Screw Threads of Unified Form |
| Pt. 2, | 1952 | Gauges for Screw Thread other than Unified Form |
| Pt. 3, | 1968 | Gauges for Screw Thread of ISO Metric Form |
| 939 | 1962 | Engineers' Squares Incl. Cyl. and Block Squares |
| 949 | 1969 | Screwing Taps |
| 957 Pt. 1, | 1941 | Feeler Gauges Inch Units |
| Pt. 2, | 1969 | Feeler Gauges Metric Units |
| 958 | 1968 | Spirit Levels for use in Precision Engineering |
| 959 | 1950 | Internal Micrometers (Incl. Stick Micrometers) |
| 969 | 1953 | Plain Limit Gauges, Limits and Tolerances |
| 1044 | 1964 | Plug, Ring and Caliper Gauges |
| 1054 | 1954 | Engineers' Comparators for External Measurement |

| BS No. | Publication Date | Title |
|---|---|---|
| 1089 | 1942 | Work Head Spindles for Grinding Machines |
| 1098 | 1967 | Jig Bushes |
| 1120 | 1943 | Diamond Tipped Boring Tools |
| 1127 | 1950 | Circular Screwing Dies |
| 1148 | 1943 | Diamond Tipped Turning Tools |
| 1580 | 1962 | Unified Screw Threads over $\frac{1}{4}$ inch Diameter |
| 1609 | 1949 | Press Tool Sets |
| 1643 | 1950 | Vernier Height Gauges |
| 1660 Pt. 1, | 1950 | Self Holding Tapers |
| Pt. 2, | 1953 | Cotter Slots and Gauges for Cotter Slots |
| Pt. 3, | 1953 | Quick Release Tapers |
| 1734 | 1951 | Micrometer Heads |
| 1759 | 1969 | Knurling Wheels |
| 1790 | 1961 | Length Bars and their Accessories |
| 1886 | 1952 | Terms and Definitions for Single Point Cutting Tools |
| 1916 Pt. 1, | 1953 | Limits and Tolerances |
| Pt. 2, | 1953 | Guide to the Selection of Fits |
| Pt. 3, | 1963 | Recommendations for Tol. Lim. and Fits for Large Diameters |
| 1919 | 1967 | Hacksaw Blades |
| 1983 | 1953 | Accuracy of Chucks for Lathes and Drilling Machines |
| 1983 Pt. 1, | 1969 | Tool Holding Chucks |
| 2485 | 1969 | Tee Slots |
| 2556 | 1954 | Hand and Breast Drills |
| 2771 | 1956 | Electrical Equipment of Machine Tools |
| 3064 | 1959 | Sine Bars and Sine Tables |
| 3066 | 1959 | Engineers' Cold Chisels |
| 3080 | 1959 | Cast Iron and Box Angle Plates |
| 3087 | 1959 | Pliers, Pincers and Nippers |
| 3088 | 1959 | Adjustable Hand Reamers |
| 3123 | 1959 | Spring Calipers and Dividers |
| 3429 | 1961 | Sizes of Drawing Sheets |
| 3616 | 1963 | Dim. of Long Milling Machine Arbors and Accessories |
| 3643 | | ISO Metric Screw Threads |
| Pt. 1, | 1963 | Data and Thread Sizes |
| Pt. 2, | 1966 | Limits and Tolerances Coarse Pitch Threads |
| Pt. 3, | 1967 | Limits and Tolerances Fine Pitch Thread |
| 3731 | 1964 | Vee Blocks |
| 3800 | 1964 | Methods for Testing the Accuracy of Machine Tools |
| 4299 | 1968 | The Overall Height of Lathe Tool Posts |
| 4311 | 1968 | Metric Gauge Blocks |

| BS No. | Publication Date | Title |
|---|---|---|
| 4372 | 1968 | Engineers' Steel Measuring Rules |
| 4442 | | Lathe Spindle Noses and Faceplates |
| Pt. 1, | 1969 | Types 'A' and 'Camlock' |
| Pt. 2, | 1969 | Bayonet Type |
| 4481 | | Bonded Abrasive Products |
| Pt. 1, | 1969 | General Features of Grinding Wheels, Blocks and Segments |
| 4500 | 1969 | ISO Metric Limits and Fits |
| 4500 A | 1969 | Data Sheet of Selected ISO Fits (Hole Basis) |
| 4500 B | 1969 | Data Sheet of Selected ISO Fits (Shaft Basis) |
| 4581 | 1970 | Dimensions of Flanges for Mounting Plain Grinding Wheels |

The above specifications may be obtained from the British Standards Institution, 2 Park Street, London, W1A 2BS, or through a bookseller.

Bona-fide students at Technical Colleges may obtain copies at reduced prices by purchasing them through their College.

# Index